Risk Principles
for Public Utility
Regulators

Risk Principles for Public Utility Regulators

Janice A. Beecher and Steven G. Kihm

Michigan State University Press | *East Lansing*

♾ The paper used in this publication meets the minimum requirements of
ANSI/NISO Z39.48-1992 (R 1997) (Permanence of Paper).

Michigan State University Press
East Lansing, Michigan 48823-5245

Printed and bound in the United States of America.

22 21 20 19 18 17 16   1 2 3 4 5 6 7 8 9 10

LIBRARY OF CONGRESS CATALOGING-IN-PUBLICATION DATA
Beecher, Janice A.
Risk principles for public utility regulators / Janice A. Beecher and Steven G. Kihm.
pages cm.—(Institute of public utilities)
Includes bibliographical references and index.
ISBN 978-1-61186-205-8 (pbk. : alk. paper)—ISBN 978-1-60917-491-0 (pdf)—
ISBN 978-1-62895-264-3 (ebook)—ISBN 978-1-62896-264-2 (kindle)
1. Public utilities—Risk management. I. Kihm, Steven G. II. Title.
HD2763.B38 2016
363.6068'1—dc23
2015023129

Book design by Charlie Sharp, Sharp Designs, Lansing, MI
Cover design by Erin Kirk New
Cover artwork © Sebastian423 | Dreamstime.com

Michigan State University Press is a member of the Green Press Initiative and is committed to developing
and encouraging ecologically responsible publishing practices. For more information about the Green
Press Initiative and the use of recycled paper in book publishing, please visit www.greenpressinitiative.org.

Visit Michigan State University Press at *www.msupress.org*

Contents

Acknowledgments

The authors gratefully acknowledge our expert reviewers: Professor Roger Buckland, University of Aberdeen; Professor Elizabeth Connors, Michigan State University; and Mr. Tommy Oliver, Virginia State Corporation Commission.

The views and opinions reflected here are those of the authors and do not represent the views and opinions of the Institute of Public Utilities, Michigan State University, or Seventhwave.

Introduction

> All courses of action are risky, so prudence is not in avoiding danger (it's impossible), but calculating risk and acting decisively.
>
> —Niccolò Machiavelli, *The Prince* (1513)

We exist in a world of scarce resources and abundant uncertainties, the combination of which can exacerbate our sense of risk. If resources were unlimited, after all, many forms of risk could be aggressively managed and largely mitigated. In reality, we must acquiesce to living with some amount of risk. Risk generally carries a negative and pessimistic connotation—that is, something "more risky" is less desirable than something "less risky." It may be typical human nature to avoid risk as well as the prospect of feeling regret. Yet, for "much of human history, risk and survival have gone hand in hand" (Damodaran 2008).

Risk rhetoric permeates contemporary regulatory discourse, and risk has become a well-worn word (Binz et al. 2012). Public utilities and regulators have always been concerned about risk because it relates directly to utilities' cost of capital, which in turn may influence allowed and earned rates of return on investment. Thus risk ultimately can affect attraction of financial capital to the utility enterprise. Unfortunately, the term "risk" can easily be misrepresented and misinterpreted, especially when disconnected from long-standing principles of corporate finance. In fact, in contrast to the conventional conception that individuals and firms should try to avoid risk, market economies and corporate cultures to a large extent actually thrive on risk because *with risk comes reward.* Not all risks can or should be eliminated, given the salient role of risk in promoting efficiency and innovation. Risk is inherent to competitive and regulated markets alike, and utilities are not exempt from risk dynamics.

Regulatory policies and practices inevitably have direct or indirect implications for allocating risk. It is unsurprising, then, that risk and risk management are prominent themes in regulatory forums. Risk today is invoked frequently as a rationale for regulatory relief in the form of higher authorized returns and modified policies and practices. Regulators should be discerning about the tendency to invoke risk as the basis for altering the regulatory paradigm, particularly by those with vested interests. Many claims about risk are thinly substantiated in terms of empirical evidence. Moreover, market actors can absorb certain risks in ways that eliminate the need for offsetting compensation. Although there is an understandable impulse to "reduce" risk, mitigating measures may be as likely to shift risk, and some might actually increase overall risk exposure. Regulators should be especially cautious about assuming risks unnecessarily by fiat, as ratepayers will likely bear the burden when risks to investors are reduced. Public utility regulators should be "risk aware" (Binz et al. 2012), but as much to maintain risk as to moderate it.

The Nature of Risk

Definitions of risk vary widely and more often than not are incomplete. Thus, confusion about risk is understandable. In common vernacular, risk is typically cast in a negative light, as in peril and the chance of incurring some form of loss. Merriam-Webster defines risk as (1) the possibility of loss or injury; (2) someone or something that creates or suggests a hazard; (3) the chance of loss or the perils to the subject matter of an insurance contract; and (4) the chance that an investment (as a stock or commodity) will lose value. In these regards, it seems natural that risk is to be avoided to the extent practicable. Individuals and businesses "insure" against risk. Organizations task personnel with "risk management," which frequently is equated with "loss management."

Regulatory advisors and analysts tend to advance the popular but asymmetric definition of risk as "the expected value of a potential loss" (Binz et al. 2012), as "exposure to a chance of injury, loss, catastrophe or any undesirable outcome" (Hoppock and Echeverri 2013), and as "the potential for a loss or negative outcome from an uncertain event" (Bean and Hoppock 2013).

As a counterpoint to colloquial and sometimes nonscientific conceptions of risk, a more technical and centric view is taken here based on the relationship of risk to expected value. Most risk is not unidirectional; that is, risky situations can present both upside and downside potential relative to expectations. Individuals and firms can choose to avoid risk, but they can also choose to embrace it. Embracing risk is rational if, within the decision maker's risk tolerance, there exists enough upside potential to outweigh the possible downside outcomes. Furthermore, although it cannot provide absolute protection, diversification can substantially limit the consequences associated with many risks.

Moreover, as we discuss throughout this treatment, risk is inherently neither desirable nor undesirable; it is merely a characteristic that informs choices. Investing in stocks is considered riskier than investing in Treasury bills, but that does not mean that one should never invest in stocks. Savers choosing to have a portion of their retirement funds invested in the stock market have overtly decided to take on risk rather than avoid it. Proper risk assessment is fundamentally about perspective. While some risks may be relevant to almost everyone, other risks may be irrelevant to all but a few. The risks that bondholders face are noticeably different from those faced by shareholders, a point that is often lost in regulatory circles. Managers face a set of risks particular to their corporate identity. Consumers are also subject to risks. Any consideration of risk is incomplete without asking the question, *risk to whom?*

Risk Principles

The objective here is to move beyond the rhetoric to a more robust and rigorous treatment of risk, particularly with regard to the consideration of risk in the regulatory context. Our approach takes a theoretically grounded but pragmatic approach to provide an integrated heuristic framework organized around well-established risk principles. Several of these might be more accurately described as corollaries, but they are treated comparably here for our purposes. The principles, and the key takeaway points from the detailed discussion of them, can be summarized as follows:

- PRINCIPLE 1. Risk is closely related to but not the same as uncertainty. Risk allows for quantitative analysis of *probabilities*, while uncertainty necessitates reliance on qualitative and often subjective assessment of *possibilities*.
- PRINCIPLE 2. Because some future outcomes cannot be envisioned, historical data are more informative about risk than uncertainty. Relying too heavily on recent experience when forecasting can lead to an overly narrow conception of possible future outcomes.
- PRINCIPLE 3. Risk is dynamic and changes with evolving conditions and new information. Risk analysis should consider the ways in which various risk factors may materialize or dissipate over time.

- PRINCIPLE 4. Risk encompasses not only the potential outcomes that are worse than expected but also those that are better than expected. Risk taking is considered rational if the potential to reap reward sufficiently offsets the potential to suffer harm.
- PRINCIPLE 5. Relevant risk is a function of covariance, that is, the variability of not one but multiple risk factors in context. Individual risk factors, no matter how consequential, should not be viewed in isolation because changes in one factor might be offset by changes in another.
- PRINCIPLE 6. Reducing risk exposure does not necessarily reduce the potential to experience regret. Regret is an emotional response based on the comparison of experienced outcomes to those that would have resulted from a different decision.
- PRINCIPLE 7. Choices and actions that minimize exposure to risk do not necessarily minimize the chance of regret (and vice versa). Decision makers should be aware of both risk and regret and the tradeoffs among these and other decision criteria.
- PRINCIPLE 8. Investors expect corporate boards and managers to maximize firm value, as reflected by stock price, not to minimize firm risk. Firm value is created by capital-deployment and operational strategies that allow companies, regulated or competitive, to earn returns in excess of their costs of capital.
- PRINCIPLE 9. Equity investors can efficiently eliminate the effect of firm-specific risks through stock portfolio diversification. Firms should manage risk on behalf of equity investors only in the unusual circumstance that firms can do so more efficiently than equity markets.
- PRINCIPLE 10. Combining risky securities in a portfolio can achieve a level of risk that is lower than that of even the lowest-risk security included. Investors benefit from diversification across and within sectors because stock price movements deviate—that is, they are not perfectly correlated.
- PRINCIPLE 11. In an efficient financial market, the only risks that matter to equity investors with diversified portfolios are systematic macroeconomic factors affecting all stocks, that is, non-diversifiable risks. Unexpected regulatory decisions are irrelevant to diversified equity investors except in the unlikely circumstance that they affect the value of all other stocks in the market.
- PRINCIPLE 12. The effects of firm-specific risk are impounded in stock prices

through cash-flow expectations, not required investor returns. A change in stock price due to any particular circumstance does not necessarily indicate a change in relevant risk to the diversified investor.

- PRINCIPLE 13. The key to risk assessment for equities is measuring the sensitivity of stock returns to changes in the value of the broad equity market. No relevant benchmark exists outside of the utilities sector for regulators to use when applying the comparable-risk standard in ratemaking.
- PRINCIPLE 14. Financial risk is a function of a firm's capital structure, as authorized for regulated utility companies. When a firm is heavily leveraged, the financial markets will require higher returns on both the debt and the equity securities issued by the firm.
- PRINCIPLE 15. A higher credit rating does not necessarily translate into a lower total cost of capital. Good reasons might exist to secure higher bond ratings by lowering the amount of debt in the capital structure, but the prospect of lowering the overall cost of capital is not likely to be among them.
- PRINCIPLE 16. Bond investors cannot easily eliminate the effect of firm-specific risks through portfolio diversification. Though relevant to bondholders, the heavy focus on firm-specific risks limits the relevance of bond-rating information to the cost of equity and equity shareholders.
- PRINCIPLE 17. Utility managers are sensitive to both systematic and firm-specific risks. Given sensitivity to firm-specific risk, corporate leadership and culture have a direct bearing on how risk is perceived and managed.
- PRINCIPLE 18. Utility ratepayers typically are captive with regard to many costs and risks, and must rely on regulators to protect their interests. Utility managers have more capacity to identify, understand, and manage various types of risk than all but the most sophisticated ratepayers.
- PRINCIPLE 19. Economic regulation and regulatory risk substitute for competition and competitive risk. It may be counterproductive to reduce risk exposure when beneficial investment, efficiency, and innovation by public utilities are desired.
- PRINCIPLE 20. Economic regulation of public utilities centers on a social compact that establishes regulatory risk and a framework for risk allocation. Reasonably allocated risk under the regulatory compact provides public utilities a path to profitability as well as essential performance incentives.

- PRINCIPLE 21. Accepted standards of regulatory review do not insulate utilities from risk or guarantee returns on investment. Regulatory risk, though substantial, is a bounded form of risk compared to the risk faced by competitive firms largely because of regulatory jurisprudence.
- PRINCIPLE 22. Fair returns are set to exceed the utility's cost of capital not to compensate for risk but to encourage socially beneficial investment. A return premium serves a social purpose because a compensatory return equal only to the cost of capital makes utilities indifferent to public policy goals attached to their performance.
- PRINCIPLE 23. All regulation is incentive regulation and all incentive regulation is based on regulatory risk. The job of the regulator is not to micromanage utilities but rather to frame the system of performance goals and incentives within which utilities must manage themselves.
- PRINCIPLE 24. Regulatory lag embeds risk in the regulatory process by design and offers both upside and downside potential. Regulatory lag should be remediated only to the extent that it substantially jeopardizes a utility's reasonable opportunity to earn a fair return.
- PRINCIPLE 25. Prudence reviews maintain regulatory risk with regard to utility investments and expenditures. Prudent behavior is expected and earns a fair return, including a reasonable but not an extraordinary return premium.
- PRINCIPLE 26. When they are deployed, incentive returns should maintain risks to motivate and reward desirable utility performance. Extraordinary return premia should be used only to promote the achievement of specific and measurable performance goals or targets.
- PRINCIPLE 27. Cost-recovery and revenue-assurance mechanisms shift risk between utility investors and utility ratepayers and thus affect the utility's overall cost of capital. Certain and expedient cost recovery and revenue assurances tend to favor investors over ratepayers and should be considered when authorizing rates of return.
- PRINCIPLE 28. Rate design can shift revenue risk between utility investors and utility ratepayers and thus affect the utility's overall cost of capital. Increased reliance on fixed over variable charges tends to favor investors over ratepayers and should be considered when authorizing rates of return.
- PRINCIPLE 29. Regulatory policies that shift risk from utility investors to utility ratepayers may increase the overall cost of service. Lower risk to

investors may come at a high price to ratepayers, namely, an offsetting loss of economic efficiency due to weak performance incentives.

- PRINCIPLE 30. Regulation provides for the periodic adjustment of rates to account for changing usage, including erosion of sales. In the regulatory context, short-term revenue risk is a function of deviations from anticipated revenue requirements and anticipated sales.
- PRINCIPLE 31. Regulation does not protect utilities from endemic economic forces and market risks, including erosion of investment opportunity. The regulatory compact and prevailing standards of review do not ensure that utilities will survive and thrive in perpetuity.
- PRINCIPLE 32. Restructuring and deregulation introduce considerable risk and uncertainty to public utilities and utility ratepayers. Competitive markets obviate the path to profitability provided to public utilities by economic regulation.
- PRINCIPLE 33. Risk should be considered as only one of many factors within a broader evaluation framework. Reducing risk exposure involves tradeoffs with other decision criteria, including costs and economic efficiency.

Ignoring risk fundamentals in the regulatory context can be perilous. Uninformed decisions may result in unintended consequences. The purpose of this treatment is to provide regulators and others in the regulatory policy community with a basic theoretical and practical grounding in risk as it relates specifically to economic regulation. This is meant as a guide not for calculating risks and returns but rather for understanding the nature of risk. The relevance of risk to regulation extends well beyond the determination of the cost of capital under the conventions of rate-of-return regulation.

Risk evokes different meanings within different disciplines. We draw from the fields of corporate finance, behavioral finance, and decision theory, as well as the broader legal and economic theories that undergird institutional economics and the economic regulatory paradigm. From a disciplinary perspective, our approach to risk actually reflects not radical but mainstream thinking. Moreover, many of the concepts explored here apply not just to financial decision making, but to decision making generally.

Our goal is to build risk literacy by challenging readers, especially economic regulators, to think intelligently, critically, and logically about risk, and perhaps even to challenge intuitive perceptions and comfort levels with regard to risk assessment. Ultimately advocates and policymakers will do what they must, but informed choices about risk (or anything else) are likely better than uninformed ones. A clearer and deeper understanding of the core concepts and principles related to risk might serve to focus and elevate the discourse around contemporary challenges in the utility sector, including potentially transformative economic, technological, and regulatory change.

Risk Principles

Risk and Uncertainty

PRINCIPLE 1. Risk is closely related to but not the same as uncertainty.

Risk allows for quantitative analysis of probabilities, while uncertainty necessitates reliance on qualitative and often subjective assessment of possibilities.

Risk is inexorably tied to uncertainty, and the terms are used interchangeably, even though the underlying concepts can be disentangled. The presence of risk means that there is some known chance that an actual result will deviate from an expected result over a particular time frame.[1] More succinctly, relative to expectations, risk "is the chance of being wrong" (Asness 2014). Risk is probabilistic, involving quantifiable estimation based on knowable information ("known unknowns"). A "risk proper" is thus a measurable uncertainty (Knight 1921) that can be understood as a set of potential future outcomes.[2] Risk manifests as discrete risk factors or simply "risks."

Classical risk assessment is focused on both the probability of an occurrence and the magnitude of its impact (table 1).[3] Normally, some amount of risk can be accepted or tolerated. High-probability high-impact events, however, clearly call for "risk management." It is useful to distinguish between endogenous risks contained within a system and exogenous risks that originate from outside forces. An endogenous risk factor is to some degree controllable.[4] An exogenous risk factor cannot be directly controlled, increased, or decreased. Attempting to eliminate exogenous risk is mostly futile. The only practical choices in the management of exogenous risks are whether to take measures or seek policies that alter exposure to risk, mitigate the effects of risk, or shift risk to others. Markets offer a variety of insurance and financial instruments toward these ends.

TABLE 1. Classical risk assessment indicating level of concern

		LOW	MODERATE-LOW	MODERATE	MODERATE-HIGH	HIGH
				IMPACT OF OCCURRENCE		
PROBABILITY OF OCCURRENCE	**HIGH**	High	High	Very high	Very high	Very high
	MODERATE-HIGH	Moderate	High	High	Very high	Very high
	MODERATE	Low	Moderate	High	Very high	Very high
	MODERATE-LOW	Low	Low	Moderate	High	Very high
	LOW	Low	Low	Moderate	High	High

Risk is sometimes considered a subcategory of uncertainty. Formally, however, uncertainty is characterized by a lack of knowledge that thwarts probabilistic measurement and estimation (including "unknown unknowns"). In life as we know it, facing uncertainty is a certainty. Uncertainty is ever-present but may or may not carry significant risk, that is, the chance of experiencing materially good or bad outcomes. Regulators themselves face uncertainty in decision making, some of which stems from the information asymmetry favoring the firms they regulate. Both risk and uncertainty are dynamic, and can grow or contract with the time frame of reference (illustrated by "fan" charts used in forecasting to show a widening range of estimated values). Although time may help resolve some risk factors as relevant information becomes known (see Principle 3), time alone does not alter risk.[5] For example, some might believe that the risk associated with investing in stocks decreases with time, that is, that risk declines the longer one holds a portfolio because gains tend to offset losses. Financial research shows that just the opposite is true (Samuelson 1994). Despite offsetting effects, the longer one holds a portfolio, the greater is the dispersion of possible portfolio values, which means that risk increases as time unfolds (Bodie 1995, 18–22).

Uncertainty is a function of the availability and accessibility of information, as well as its quality or "ambiguity." When they share "perfect" information, all parties will agree on the likelihood of all future outcomes. In reality, agreement about probable outcome is unlikely in part because of uneven and unclear information. Ambiguity arises from problems of ignorance or circumstances that are inherently information-poor, and because acquiring information can be costly. Because any individual's knowledge is incomplete, different persons can rationally draw different inferences about the likelihood of particular results. For example, ambiguity about the pace of climate change may affect perceptions of urgency with regard to policy response. Fortunately, for some problems, scientific research can help overcome ignorance and reduce ambiguity.

Behavioral finance and decision analysis recognize the difference between risk aversion and ambiguity aversion. Because of its probabilistic nature, risk may be easier to accept than uncertainty, particularly when the problem of ambiguity is substantial. Without effective tools for coping with imperfect information, uncertainty can stifle decision-making processes. Uncertainty is further exacerbated by "complication" (many factors) and "complexity" (dynamic and adaptive interaction among factors). For heuristic purposes (including ours), it often is necessary to hold some factors constant in order to understand risk mechanics. Advanced theories and modeling methods are aimed at overcoming the "analysis paralysis" associated with the challenges of uncertainty in a complicated and complex world.

Perhaps because it seems more tangible, the term "risk" is invoked frequently when more often than not the term "uncertainty" would be technically correct. Risk allows for quantitative analysis of *probabilities,* while uncertainty necessitates reliance on qualitative and often subjective assessment of *possibilities.* Highly uncertain situations force greater reliance on informed intuition, which is likely shaped by experiential and cognitive biases.

Risk and Future Outcomes

PRINCIPLE 2. **Because some future outcomes cannot be envisioned, historical data are more informative about risk than uncertainty.**

Relying too heavily on recent experience when forecasting can lead to an overly narrow conception of possible future outcomes.

The major mistake associated with viewing uncertain situations through a risk-based lens is the false belief that the range of possible outcomes is known, or at least knowable, in advance. Processing uncertainty can be a substantial cognitive challenge; even highly creative individuals will fail to envision the full gamut of possibilities.

Instead of pushing the limits of imagination, humans tend instead to succumb to "recency bias." Relying too heavily on recent experience when forecasting can lead to an overly narrow conception of possible future outcomes. The oil industry provides a classic example of this difficulty.

In the period following World War II and until the 1970s, oil prices were relatively stable (figure 1). In the late 1960s, analysts forecasting oil prices did not see, and reasonably could not have been expected to see, the drastic changes that the industry would undergo in the coming decade, including the Iranian Revolution and two OPEC-led embargoes. Yet, these "unknowables" are what uncertainty is all about. Contemporary experience also makes the point that predicting oil prices has not become easier with time.

An economic regulatory example further reinforces this concept. Consider a utility's plan in the early 1970s to build a nuclear power plant at a projected cost of $1.3 billion. Long-term construction projects increase sensitivity to systematic economic

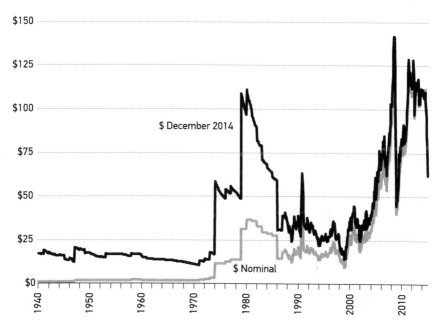

FIGURE 1. Oil prices (Europe Brent Spot Price), 1940–2014

SOURCES: DATA FROM THE *BP STATISTICAL REVIEW OF WORLD ENERGY*, JUNE 2014, AND THE U.S. ENERGY INFORMATION ADMINISTRATION (WWW.EIA.GOV).

risks (see Principle 11). Unexpectedly, the utility encountered substantial midstream construction delays as well as record-high interest rates when it needed to borrow billions of dollars more than planned to complete the plant. By the time the plant came on line in the late 1980s, construction costs had more than tripled. Because the utility could not obtain the rate relief necessary to fully recover the plant's cost, it sought bankruptcy protection (a rare event in the sector). It is unlikely that any investor risk assessment conducted in 1972 envisioned that possibility (Guthrie 2006).

Of course, overly simplistic methods of forecasting are more likely to suffer from recency bias as well as problems of obsolete or unrepresentative data. When based on well-informed and realistic probability distributions, modern simulation methods offer substantial analytical improvement in terms of addressing data challenges and modeling the consequences of unusual and uncommon behaviors and events, including "black swans" (see Taleb 2010).

Risk Dynamics

PRINCIPLE 3. **Risk is dynamic and changes with evolving conditions and new information.**

Risk analysis should consider the ways in which various risk factors may materialize or dissipate over time.

Although frequently treated as static, risk itself is dynamic. Risk and associated impacts can change dramatically over time as some outcomes become known. Conditions of uncertainty, in other words, can evolve into quantifiable risk, and risk probabilities can change with changing circumstances that affect various risk factors. As some risks and uncertainties expire, others may emerge. Risk analysis should consider the ways in which various risk factors may materialize or dissipate over time.

In project management, the "cone of certainty" describes how various material issues are resolved successively as a particular project is actually implemented (figure 2). With each stage of a project, decision makers benefit from actual experience (that is, the realization of risks one way or another). At the same time, however, decisions can become constrained or path-dependent due to prior approvals and sunk costs. For this reason, flexible design, incrementalism, modularization, pilot programs, and phase-in plans can be effective strategies for project-related risk *management* (as compared to risk avoidance). For capital-intensive utilities, the potential advantages of these tools must be weighed against the economies of scale in construction and operation traditionally associated with large or "lumpy" capital investments.

To illustrate the dynamic nature of risk, consider a firm that rents office space but pays the energy bills. An energy audit reveals that the firm could implement a

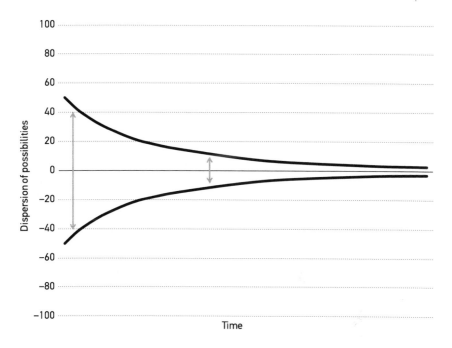

FIGURE 2. Project uncertainty over time (hypothetical data)

proven efficiency technology that will reduce its energy bills by 10 percent with a payback period of three years. The rate of return on the project over the next ten years is 18 percent.

The office manager knows, however, that the building lease will expire in six months, and there is a 75 percent chance that the lease will be renewed for ten years and a 25 percent chance that the firm will relocate. Making the energy-efficiency investment prior to the lease decision is quite risky and probably unwise. If the firm does not renew the lease, it would leave before accruing most of the benefits of the improvements. The situation changes, however, once the lease decision is made. If the firm chooses to stay, the investment is a low-risk way of producing an attractive return. In this case, it is the decision to stay or move, not the characteristics of the efficiency technology, that creates the relevant risk. Once the lease decision is made, there exists little remaining risk, as should be reflected in the analysis.

Analyzing projects with dynamic risk characteristics has perplexed financial experts for decades. The discounted cash-flow model, which forms the backbone of traditional financial analysis, offers a largely static perspective of risk. Although complex and not widely applied in the utilities sector, real-options pricing theory addresses the challenge of modeling risk dynamics.[6]

The important message is that some forms of risk may not remain constant over the course of a project or decision period. Because some risks might come or go, decision makers should ask whether mid-course information has the potential to affect risk and its relevance. When risk is dynamic, risk analysis becomes more complicated and simplistic analytical tools might provide incomplete or misleading answers.

Risk and Opportunity

PRINCIPLE 4. **Risk encompasses not only the potential outcomes that are worse than expected but also those that are better than expected.**

Risk taking is considered rational if the potential to reap reward sufficiently offsets the potential to suffer harm.

Importantly, the presence of risk or uncertainty does not mean that venturing forth along new paths is to be avoided. Despite the human inclination to view risk in the negative, the well-known adage "nothing ventured, nothing gained" points to the positive side of risk. In business, as in other walks of life, where there is risk there is opportunity (Damodaran 2011). Likewise, risk avoidance may incur opportunity costs. Indeed, identifying opportunities and taking calculated risks based on established principles might be the key to competitive advantage and corporate success.[7]

Finance theory treats most risks as broadly symmetrical and randomly dispersed (or normally distributed). Although not always perfectly so (due to skewing and clustering), downside risk is generally balanced by upside risk. For businesses, upside risk suggests a more positive connection between risk and firm performance, including both efficiency and innovation. Whether competitive or regulated, effective enterprises are able to recognize and manage risk in ways that enhance firm value and returns.

Corporate risk management often leads firms seeking to increase investor value to take certain risks rather than avoid them. As one CFO explained in a *Harvard Business Review* interview: "The route to success is to put *more* money at risk, not less" (Nichols 1994). Corporate finance principles go further to suggest that taking

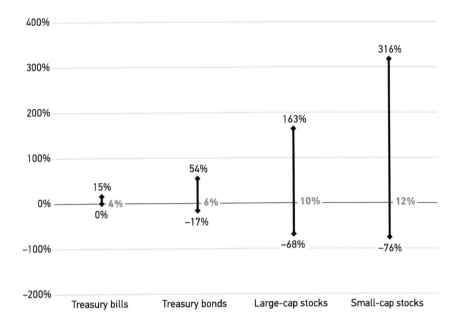

FIGURE 3. Range of annual returns by security type (highest, lowest, and average), 1926–2010

SOURCE: DATA FROM DAMODARAN ONLINE (HTTP://PAGES.STERN.NYU.EDU/~ADAMODAR).

action solely to reduce risk exposure is a waste of both time and money (Brealey et al. 2006, 721). Thus many conventional definitions and uses of the term "risk" do not hold sway inside the corporate finance world.

Seasoned investors understand the association of risk and return over various investment vehicles and time horizons. Investors looking to earn higher annual returns over the long run can do so only by taking on risk, that is, by embracing both downside and upside potential. It is the upside potential that can make risk-taking attractive under some circumstances.

Evidence suggests that financial markets price securities so that investors are compensated for taking on risk commensurate with their expectations about returns. The disparity or range of returns is used as a simple measure of risk, as shown in figure 3. For large-cap stocks, for example, the difference between the highest return year (163 percent) and the lowest return year (–68 percent) is 231 percentage points for the period analyzed. For Treasury bills, the difference between the highest annual return (15 percent) and the lowest (0 percent) is noticeably lower, reflecting the low risk associated with this type of security.

Whether an individual investor wishes to take on the higher risk of investing in stocks in the hope of earning potentially higher returns is a matter of personal risk tolerance. For any decision, the risk-taker expects to be appropriately rewarded. Risk taking is considered rational if the potential to reap reward sufficiently offsets the potential to suffer harm. Likewise, risk taking is considered irrational if the potential to suffer harm is not sufficiently offset by the potential to reap reward.

Risk as Covariance

PRINCIPLE 5. Relevant risk is a function of covariance, that is, the variability of not one but multiple risk factors in context.

Individual risk factors, no matter how consequential, should not be viewed in isolation because changes in one factor might be offset by changes in another.

Risk is commonly associated with the concepts of variability or volatility, understood as a dispersion of values (such as investment returns over a period of time). For example, because natural-gas prices regularly move up and down, they are considered "risky." This conceptualization equates risk with statistical variance (typically measured in terms of standard deviation) and can lead people to overstate certain risks. Technically accurate risk assessment focuses not on the *variance* for any single variable but on *covariance*, or how a set of relevant variables moves over time (D'Ambrosio 1990, 2–18). This approach also recognizes that risk is both complex and dynamic and that these attributes are relevant to risk assessment and response.

Consider, for example, whether tighter emission regulations might cause natural-gas prices to rise as more utilities drive up demand by turning to gas instead of coal for generation. As fuel is a major input to the power production process, this would be a bad outcome for some utilities and their ratepayers. Nevertheless, other utilities (namely, those with more modern and cleaner coal-burning plants) might be well prepared to meet standards and thus benefit from falling coal prices, as coal falls out of favor as a fuel choice.

Both natural-gas prices and coal prices manifest variance, but the key question is whether the associated fuel-price risks are relevant to all utilities. The answer is no,

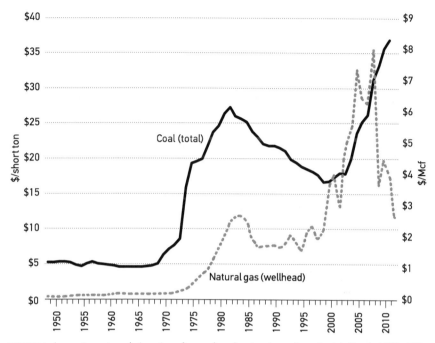

FIGURE 4. Long-term trends in prices for coal and natural gas (nominal dollars), 1949–2011
SOURCE: DATA FROM THE U.S. ENERGY INFORMATION ADMINISTRATION (WWW.EIA.GOV).

at least not to the same degree, because while the simultaneous price movements of the two commodities were positively related in the past, more recently they have moved in opposing directions (figure 4). While a utility that relies exclusively on one generating fuel type will find itself in a risky position, a utility operating a mix of coal-fired and gas-fired plants may see limited changes in aggregate fuel costs. Relevant risk is affected by the degree to which risk factors are nonrandom or correlated. This is the essence of resource diversification.

Thus, whether fuel-price volatility represents relevant risk to any particular utility depends on the composition of the utility's resource portfolio and regulatory policy as well as whether cash-flow variability associated with fuel prices correlates with changes in the broader equity market.[8] In other words, individual risk factors, no matter how consequential, should not be viewed in isolation because changes in one factor might be offset by changes in another. Understanding these portfolio effects is what covariance and relevant risk are all about. These concepts apply to all risk factors, not just fuel prices.

Risk and Regret

PRINCIPLE 6. Reducing risk exposure does not necessarily reduce the potential to experience regret.

Regret is an emotional response based on the comparison of experienced outcomes to those that would have resulted from a different decision.

Taking risk can lead to a negative outcome. But avoiding risk altogether in some instances guarantees dissatisfaction. Lord Tennyson's "'Tis better to have loved and lost than never to have loved at all" is the literary version of this axiom. One can avoid the potential pain and joy associated with a loving relationship by never having one, that is, by never taking a risk in this regard. What Tennyson's poem conveys is that avoidance behavior will ultimately lead to regret for not having even tried to pursue a loving relationship.

Regret is an emotional response based on the comparison of experienced outcomes to those that would have resulted from a different decision (also known simply as second-guessing). For human beings, experiencing regret is natural and inevitable. Fear of regret and the lingering doubt about making a "bad" choice are familiar to many if not most financial investors. New investors are often asked to complete a "risk tolerance" assessment, which may be as much about "regret tendency." Research suggests that the sense of disappointment from loss can be more pronounced than the sense of satisfaction from gains, and this disparity can shape investment behavior (Richards 2013). A risk-averse investor might choose a portfolio consisting only of Treasury securities. While the lack of volatility might provide comfort, especially during market downturns, investors in their twilight years might feel substantial regret about the end result, not because they took on too much risk but because they did not take on enough. Their realized returns could

TABLE 2. Risk and regret under gas price risk hedging by customers or utilities

| | PRICE RISK | EXPERIENCED OUTCOME | |
		GAS PRICE BELOW HEDGED PRICE	GAS PRICE ABOVE HEDGED PRICE
Customers or utility does not hedge	Yes	Customers and regulators experience no regret	Customers and regulators experience regret
Customers or utility hedges	No	Customers and regulators experience regret	Customers and regulators experience no regret

Note: The capacity for hedging by regulators or customers depends in part on market structure. In deregulated markets, customers alone make hedging choices.

have been much greater if more risk had been accepted. In economic terms, the risk-averse investor incurs an *opportunity loss.*

Even diversified investors sense some regret when they fail to "buy more at the bottom" or "sell more at the top" of market cycles. These are the "could have/would have/should have" dilemmas of investment and life. Sophisticated investors tend to not experience substantial regret, because they understand that investing is not a game and that timing the market is virtually impossible. This does not mean that investing is less risky for these investors; it just means that they process risk more rationally and less emotionally.

Thus, risk and regret are distinct concepts with distinct roles in behavioral finance and decision analysis. Risk is an *ex ante* decision consideration reflecting the likelihood of particular outcomes. Regret is an *ex post* emotional response based on the comparison of experienced outcomes to those that would have resulted from a different decision (that is, a counterfactual scenario).

While regret will be felt retrospectively, it can be considered prospectively when making choices. Thus, regrets analysis can be used for making decisions in the face of uncertainty, when insufficient or ambiguous information thwarts probabilistic estimation of outcomes. The potential for regret can motivate risk-avoiding or risk-taking behavior. In some respects, risk relates to *investment* and regret relates to *insurance.* Insurance products, including extended warranties and annuities, are meant to minimize regret associated with a catastrophic loss (just as they might also lead to some regret associated with opportunity costs if a loss circumstance does not materialize).

A public-utility example of the difference between risk and regret can be found in fuel-price hedging (table 2). If a utility uses future contracts to hedge gas prices over the coming heating season, it reduces gas-price risk. The fact that gas prices might decline over the coming year, causing ratepayers to pay more for gas than they otherwise would have, is not a risk associated with hedging; it represents the potential for utility managers, ratepayers, and regulators to feel regret.

There is no way to avoid the potential for feeling regret in a hedging context. Hedging prior to a price decrease will lead to regret for having done so; failing to hedge prior to a price increase will lead to regret for not having done so. The potential for regret therefore exists in both the hedged and the unhedged positions, but this does not bear on the risk associated with either choice. The unhedged position is risky because the price to be paid over the heating season is uncertain. The hedged position is not risky because a utility knows what it will pay for gas over the heating season. In other words, hedging can help participants in the market economy cope with risk but not regret.

Importantly, hedging for one factor reduces only exposure to that one source of risk, not all risk. As noted, overall risk is a function of covariance. To the extent that one variable is negatively correlated with another, hedging for one variable but not the other could actually make a portfolio more risky. The choice to hedge in one area might also constrain other managerial choices affecting risk and performance.

Risk Minimization

PRINCIPLE 7. Choices and actions that minimize exposure to risk do not necessarily minimize the chance of regret (and vice versa).

Decision makers should be aware of both risk and regret and the tradeoffs among these and other decision criteria.

The hedging example discussed in the previous section shows that the choice to either embrace or eliminate fuel-price risk does not eliminate the chance of regret. It follows that for planning and decision-making purposes, analyzing risk is not the same as analyzing regret or, for that matter, other relevant criteria. In fact, the goals of risk and regret minimization can be at odds.

Focusing on the potential for regret can bias decision analysis. The "minimax" criterion, as implied, focuses on the goal of minimizing the maximum prospect of regret. Regrets analysis can be informative but overly conservative if it emphasizes only the worst possibilities, regardless of likelihood. A full regrets analysis considers the full range of potential outcomes, good and bad.

Confusing risk and regret, and hybrids of the two concepts, confounds decision analysis. Some analysts mistakenly treat risk and regret either as synonymous or as correlated decision criteria.[9] A basic investment scenario is illustrated in figure 5. When given a choice between investment alternatives, the least-risk portfolio clearly is one consisting of all Treasury bills. Risk depends on year-to-year variability in returns, which is quite low in this instance. But regret, which is experienced after the fact, depends on the difference between what was earned with the all-Treasuries portfolio and a counterfactual case, that is, what might have been earned with a different, higher-risk portfolio.

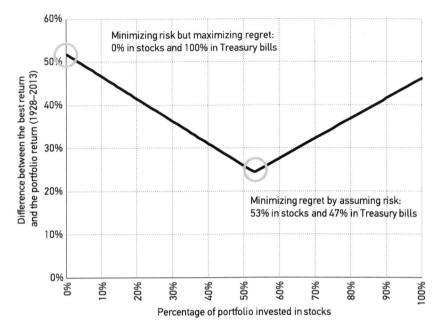

FIGURE 5. Minimizing risk vs. minimizing regret in investing, 1928–2013

SOURCE: DATA FROM DAMODARAN ONLINE (HTTP://PAGES.STERN.NYU.EDU/~ADAMODAR).

For the period 1928–2013, the maximum annual regret under an all-Treasuries portfolio materialized in 1954, when stocks increased by 53 percent while Treasuries rose by only 1 percent. Conservative investors missed an opportunity to earn an additional 52 percentage points of return that year. Investors who hold Treasury bills never lose money, but neither do they make much money compared to equity investors; that is, insulation from market volatility comes at a price. Over the long run, investors who rely exclusively on Treasury bills forego, on average, 6 percentage points per year in investment returns (relative to shareholders of common stocks).[10] Retirees who took on risky investments in their working years typically accumulate far more in their portfolios than those who played it safe. Whether this will be the case going forward is not the point, since regret is measured after the fact, but those who missed out on the golden age of investing that was the twentieth century no doubt experienced considerable regret for choosing comfort over opportunity. In other words, low risk can lead to high regret.

Analysis shows that over this same historical period, the least-regret portfolio (that is, the one with the lowest maximum difference between the return on the investor's portfolio and that of the best-performing portfolio) consists of 53 percent stocks and 47 percent Treasury bills. No one would argue that this mix represents the lowest-risk option, which is the all-Treasury bill portfolio. When applied properly, the minimax criterion actually leads to a diversified mix of Treasuries and stock securities. Although in the abstract a scenario where both risk and regret are minimized is conceivable, the illustration dispels the proposition that these objectives are one and the same.

Thus, typical investment portfolios do not minimize risk in an absolute sense; they limit unnecessary exposure to risk given a particular level of risk tolerance as well as an appreciation of the potential for regret. Even the most diversified stock portfolio is much riskier than investing in Treasury bills alone. Investors include stocks in their portfolios because they view risk as offering both upside and downside potential. Their goal is not to minimize risk, but to maximize returns for the level of risk they are willing to take. These disciplined investors experience little regret. In general, decision makers should be aware of both risk and regret and the tradeoffs among these and other decision criteria. Risk-aware decision makers should be willing to accept some risk in order to minimize the potential to experience regret.

Risk and Firm Value

PRINCIPLE 8. Investors expect corporate boards and managers to maximize firm value, as reflected by stock price, not to minimize firm risk.

Firm value is created by capital-deployment and operational strategies that allow companies, regulated or competitive, to earn returns in excess of their costs of capital.

It goes without saying that investors expect a return of and on their investment, reflected in interest payments and dividends as well as security prices. Savvy investors understand the connection between risk and returns, and in fact, compare investment choices on the basis of risk-adjusted returns and time-sensitive expectations.

Companies, like individuals, can choose to avoid exposure to risk or chance of regret. But as shown previously, risk avoidance can result in substantial opportunity losses. Investors buy stocks, instead of Treasury bills, with the understanding that equity returns can be realized only by accepting some amount of risk. Over the long run, returns tend to suffer from risk avoidance. Corporate boards and executive teams are obligated to investors with respect to maintaining and enhancing the total value of their investment over time. In reality, most shareholders expect boards and managers to take calculated and strategic risks in order to enhance total returns, and hold them accountable for doing so.

An obvious theoretical and practical question is, how do boards and managers create value for their investors? Simply put, firm value is created by capital-deployment and operational strategies that allow companies, regulated or

TABLE 3. Project risks and returns relative to the cost of capital and investment scale (hypothetical data)

JURISDICTION	PROJECT RISK	COST OF EQUITY	EXPECTED RETURN ON EQUITY	EXCESS RETURN TO INVESTOR	INVESTMENT SCALE	NET PRESENT VALUE
1	Low	8.0%	11%	3.0%	$500,000	$117,139
2	Moderate	9.5%	10%	0.5%	$4,000,000	✓ $141,511
3	High	10.0%	11%	1.0%	$1,000,000	$68,577

competitive, to earn returns in excess of their costs of capital. Neither maximizing returns nor minimizing risk necessarily satisfies the total-value obligation.

When capital is limited, firms may need to choose among projects with different risk profiles. For a plausible example, consider a firm that is contemplating project options in three different jurisdictions, each with a different strategic risk profile (table 3). Assuming that the lifetimes for all three projects are the same, the most valuable project in this example is the one in the second jurisdiction, which provides the lowest return on equity coupled with a moderate risk profile. In this case, expected returns exceed the cost of equity by a relatively small differential, but project *scale* is determinative of net present value.[11] The moderate-risk option creates the most value for investors, even though the investment option neither maximizes returns nor minimizes risk. Most investors would expect corporate managers to act accordingly. In other words, given the goal to maximize value, firms that pass up on investments in "low-return" jurisdictions, based on that decision criterion alone, might forgo gains for their shareholders.

This type of analysis can be extended to a wide range of strategic issues. Most corporate decisions about expansion, as well as mergers and acquisitions, are likely evaluated in terms of effects on firm value; certainly the investment community takes this perspective. Along these lines, corporate growth may or may not be desirable in terms of potential value creation and may affect investors of acquiring and acquired companies differently.[12] In reality, of course, strategic decisions for individual firms are made within a complex set of economic and regulatory considerations.

Risk and Equity Investors

PRINCIPLE 9. Equity investors can efficiently eliminate the effect of firm-specific risks through stock portfolio diversification.

Firms should manage risk on behalf of equity investors only in the unusual circumstance that firms can do so more efficiently than equity markets.

For equity investors, risk is manifested in the range of possible returns; the wider the range of returns, the greater the risk. As illustrated in figure 6, holding multiple securities that are diversified within and across investment sectors or asset classes (manufacturing, technology, health care, etc.) is key to managing portfolio risk. Firms themselves may venture into various investment activities; that is, they may attempt to diversify on behalf of their shareholders. The prevailing finance literature has long held, however, that firms should manage risk on behalf of equity investors only in the unusual circumstance that firms can do so more efficiently than equity markets (see Brealey and Myers 1984).

Not all equity investors and not all utility shareholders are diversified. Yet this alone is not a compelling reason for regulators to consider firm-specific (nonsystematic) risks when estimating a utility's cost of equity or authorizing rates of return. Failing to diversify risk suggests that the investor lacks financial sophistication or seeks investment risk (perhaps predicated on a sense of special advantage in this more risky situation).[13] Of course, the non-diversified investor or manager will find little comfort in the idea that diversified investors are unaffected by firm-specific events, nor will the utility regulator, who might face political consequences of lower credit ratings or stock prices for affected companies. Nonetheless, ratepayers should not be responsible for compensating investors for

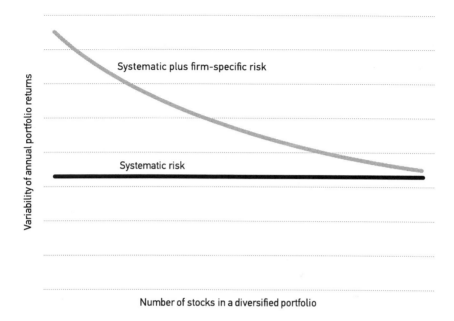

FIGURE 6. Management of firm-specific risk through diversification (hypothetical data)

poor investing practices or satisfying the tastes of investors who prefer to live on the edge with respect to risk.[14]

Most importantly, even though many utility investors might not be diversified, they are not likely significant players in the overall equity market. Diversified institutional investors dominate trading in securities, both in terms of scale of investment and frequency of trading, and they are the only market participants that provide price signals in terms of risk assessment and required returns (that is, the cost of equity). It is these dominant investors, who own a significant equity stake and trade shares at the margin, that influence equity prices (Damodaran 2011, 59).

Utility managers are appropriately sensitive to how risk affects all of their equity investors. However, most individual equity investors own too few shares to influence stock prices even if they trade frequently. Furthermore, individuals tend to be "buy-and-hold" investors who do not actively trade stocks. As such, individual investors are "price-takers" who have essentially no influence on either the cost of equity or the price of stocks, making their perceptions of risk largely irrelevant to these determinations. So even if a utility has a large proportion of non-diversified investors, the risk-return concerns of this group are not revealed. In sum, the risks that matter are those perceived by the institutional investors that move the equity markets.

Risk and Diversification

PRINCIPLE 10. **Combining risky securities in a portfolio can achieve a level of risk that is lower than that of even the lowest-risk security included.**

Investors benefit from diversification across and within sectors because stock price movements deviate—that is, they are not perfectly correlated.

A core principle of finance is that holding many securities in a portfolio, rather than owning a single stock or bond, can dilute and mitigate the effects of risk (Principle 9). The Dow Jones Utility Index represents a portfolio consisting of fifteen utility stocks that exhibited little price appreciation from the early 1970s through the mid-1980s, but since that time the price trajectory has clearly been upward-sloping. For the past twenty-five years, utility stock prices have increased at an average rate of about 4 percent per year. Utility dividend yields averaged about 6 percent over that period, producing a total annual return of nearly 10 percent, allowing typical utility investors to double their investment value every seven years.

That is not to say that all utilities fared well over this period; although an unusual occurrence for the sector, three major electric utility companies and one major gas utility company filed for bankruptcy protection during that time.[15] A utility investor diversified across a wide number of companies and industries would be largely unaffected by such extreme but firm-specific events. If the portfolio contained hundreds of stocks for various sectors, returns on the other stocks in the portfolio would have more than offset the bankruptcy-related losses. To be sure, the effect of bankruptcy on a firm, its management, and its non-diversified investor might be devastating, but the effect on a well-diversified portfolio is usually *de minimus*.

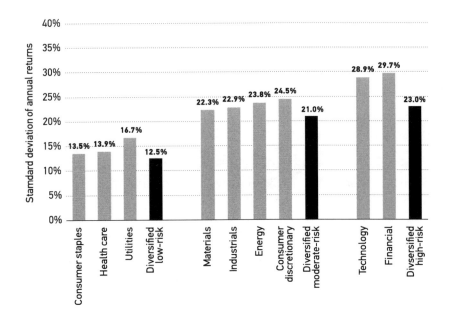

FIGURE 7. Effects of diversification on annual returns by sector, 1999–2012

SOURCE: DATA FROM ALTAVISTA, *ETF ANALYZER: SELECT SECTOR SPDR EDITION*, MARCH 2014.

This portfolio effect permeates financial markets. Figure 7 shows the standard deviation of the annual returns[16] for nine sector index funds ranked from lowest standard deviation (lowest risk) to highest standard deviation (highest risk). Data on the variability of stock returns (standard deviations) suggest that the sectors cluster into three risk groups: low, moderate, and high. The data also reveal the benefits of diversification both across and within investment sectors. As discussed under Principle 5, what undergirds the notion of relevant risk is the covariance across firms and sectors, not the variance of either the firm or the sector alone.

Investors preferring lower-risk securities might select the utility fund. Alternatively, they could choose to invest in all three of the lower-risk sectors, with the seemingly paradoxical result of a portfolio with lower risk than any of the three component sectors. Investors benefit from diversification across and within sectors because stock price movements deviate—that is, they are not perfectly correlated.[17] In other words, rather than moving directionally in lockstep, one fund's upside movement tends to offset another fund's downside movement. Both contribute toward the standard deviation (or variance) of the individual sector fund, but their respective effects are canceled out in the diversified portfolio. Thus, combining relatively high-risk securities can actually produce a relatively low-risk portfolio. Portfolio risk can be substantially lower than that of the individual components if the holdings are not significantly correlated. This is the logic behind portfolio and mutual-fund investing.

Consistent with modern portfolio theory and optimizing returns relative to risk (Jones 2010, 173), the cross-sector analysis shows that diversification produces powerful nonlinear effects. The risk associated with a portfolio of companies is not simply the average of the standard deviations of the individual components. Rather, the standard deviation of a diversified portfolio that combines securities is often lower than the standard deviation of the lowest-risk component in the group. Further diversification can be achieved through a mix of equity (stock) and debt (bond) investments along an efficient frontier with respect to the relationship of risk to expected returns. Differences in risks for stock and bond investing are discussed later (see Principles 11 to 16).

Relevant Risk

PRINCIPLE 11. In an efficient financial market, the only risks that matter to equity investors with diversified portfolios are systematic macroeconomic factors affecting all stocks, that is, non-diversifiable risks.

Unexpected regulatory decisions are irrelevant to diversified equity investors except in the unlikely circumstance that they affect the value of all other stocks in the market.

Portfolio diversification can be "perfect" when the investor holds the stocks of both the firm that gains and the firm that loses subsequent to some occurrence. If Company A's technology captures some of Company B's market share, Company A's stock goes up, while Company B's stock goes down. Holding both companies eliminates risk associated with this form of competition.[18] Perfect matching of firms in a portfolio, however, is not necessary to achieve risk reduction. Since any firm-specific risk affects at best a handful of firms, a random selection of stocks will provide the same sort of protection from those risks. For example, holding Company C from another sector in the portfolio would also offset the decline in Company B stock, even though the two firms do not compete directly in the same market. Similarly, if production costs rise for Company D, Company E might stand to gain if it supplies inputs to Company D.

The diversification principle does not require all or even most investors to be well diversified. As noted, individual investors who buy and hold utility stocks have no impact on the cost of equity. The well-diversified institutional investors that hold many funds and buy and sell large blocks of stocks set the stock price, and therefore the cost of equity. The market will price stocks to reflect only non-

TABLE 4. Select manifestations of risk: Systematic, sector-specific, and firm-specific

SYSTEMATIC RISKS	SECTOR-SPECIFIC RISKS
▪ Fiscal and monetary policy	▪ Sweeping reforms or regulations
▪ Persistent economic recession	▪ Financial reporting requirements
▪ Inflation, deflation, or stagflation	▪ Taxes on carbon emissions
▪ Change in currency valuation	▪ A change in depreciation rules
▪ Widespread war or epidemic	▪ Accounting treatment of income taxes
▪ Catastrophic climate change	▪ A new technical standard
	▪ Disruptive technologies
	▪ Change in consumer preferences
	▪ Disturbance in commodity markets
	▪ Pension-fund requirements

FIRM-SPECIFIC RISKS

- Failure to secure a patent on a new technology (or success in obtaining one)
- A looming strike by employees (or an unexpected successful labor contract result)
- The retirement of an iconic chief executive officer (or the hiring of a talented new one)
- A cyber-security violation (or recovery of funds stolen by hackers)
- Weather effects that dampen customer demand (or increase it)
- An unexpected infrastructure failure (or plant lasting longer than expected)
- A large-volume customer leaves (or one moves into the service area)
- A construction project results in cost over-runs (or a project comes in under budget)
- Corporate bankruptcy (or an unexpected recovery that avoids it)
- A change in environmental rules that harm a firm (or make it more attractive)
- A cost disallowance by regulators (or an unexpected decision to allow cost recovery)
- A lower than expected authorized return (or one that is higher than expected)

diversifiable risks because diversified investors will pay more for a stock than will non-diversified investors; non-diversified investors are price-takers in this context.

Diversification clearly is helpful in managing risk, but it cannot protect investors from the systematic risks affecting all stocks at once. When systematic risk factors hit the market, all boats—that is, all firms—rise or fall with the changing tide. The tide analogy is not quite perfect in that while all boats rise or fall with the tide to the same extent, the degree to which individual stock prices rise or fall with broad equity market changes can vary. It is not firm-specific risks but the *sensitivity* of the firm's stock to these market changes that matters to equity investors.

The one-year period between March 3, 2008, and March 2, 2009, when the value of the stock market declined by almost 50 percent, is illustrative. Although no investment sector was entirely immune from loss, the impact varied across sectors.[19] This massive, highly correlated, system-wide shock (also referred to as a "black swan" event) is the scenario that investors fear the most because it is the one in which there is nowhere for even the most skilled investor to hide within the stock market.[20]

Thus, to fully appreciate the consequences of risk-taking, utility executives and regulators must distinguish between the systematic or market risks that matter to diversified equity investors and the firm-specific, business, or idiosyncratic risks that do not; in between are sector-specific risks that matter under some circumstances (table 4).

Systematic risks that are relevant to investors include unexpected macroeconomic or global geopolitical events that cut across all sectors of the economy. Systematic risk is threatening because as the cost of equity goes up, even if earnings and cash flow remain the same, all stock prices decline. Even a well-diversified portfolio provides little protection from macro-level or systematic risk factors that portend market instability or even collapse.

Sector-specific risks affect groups of like investment securities (such as electric utilities) within the overall market in like ways (that is, nonrandomly). To the extent these risks can be managed through portfolio diversification, they are irrelevant. To the extent that they differentiate the sector in terms of variation in expected returns, they may reflect sector sensitivity to systematic risk. In other words, sector-specific risks might be relevant to investors in terms of asset allocation if their effects ripple across the economy.

The list of firm-specific risks is interminable yet irrelevant to equity investors because these risks are random and diversifiable. Corporate managers naturally think of the many things that can go wrong in the course of time, including failure to secure regulatory approvals. However, these risks should not be used to rationalize the use of "fudge factors" in estimating required returns (Brealey et al. 2006, 223). Senior investment advisors acknowledge that since most of the risks that companies face are in fact diversifiable, most risks do not affect their cost of equity capital (Koller et al. 2010, 34). Effective portfolio managers thus ignore idiosyncratic risks and focus only on the portfolio's sensitivity to "a relatively small list of macro variables and their associated implications for maximum efficiency" (Diermeier 1990, chapter 5, page 2).

Regulatory risk (discussed in detail under Principle 19) should generally be understood as a *firm-specific risk*. An important clarification is that a firm-specific regulatory action might increase the utility's exposure or sensitivity to systematic risks. For example, if a regulator eliminates returns on construction-work-in-progress (CWIP) for a new power plant, the company will turn to the financial markets for additional capital. Changes in interest rates (a macroeconomic risk) will affect all companies in the economy, including utilities with respect to capitalized construction costs. Thus, an unexpected change in interest rates (that is, interest-rate risk) is a relevant risk for the utility's equity investors. Sensitivity to risk is a function of participation in the broader economy and also relative; that is, bigger companies and bigger projects mean more exposure.

Understandably, regulators can be sensitive to the effect of their decisions on utility stock prices. For the most part, however, regulatory risk related to cost recovery or authorized returns for a particular utility has no effect on other companies (regulated and non-regulated). Even though it will affect the utility's stock price (by affecting cash flow), the risk is diversifiable to equity investors. Thus, regulators should be circumspect about weighing financial market reactions to their decisions. Indeed, the well-cited *Federal Power Commission v. Hope Natural Gas Co.* case reinforces the point that the regulator's job is not to maintain or enhance the current value of stock prices but to make decisions that are *in the public interest.* As the Supreme Court found:

> The fixing of prices, like other applications of the police power, may reduce the value of the property which is being regulated. The fact that the value is reduced does not mean that the regulation is invalid. . . . The heart of the matter is that rates cannot be made to depend upon "fair value" when the value of the going enterprise depends on earnings under whatever rates may be anticipated.[21]

Failing to distinguish types of risks can lead to flawed financial analysis and an overstatement of required investor returns. The corporate finance literature hammers home these points. Diversified equity investors face only a few relevant risks, and these are not specific to any particular firm, including a regulated utility.

In financial markets, taking on risk leads to higher expected returns only when the risk is relevant to the equity investor. Systematic risk affects a firm's cost of equity; firm-specific risk does not. The implications for regulated returns are important. Unexpected regulatory decisions are irrelevant to diversified equity investors except in the unlikely circumstance that they affect the value of all other stocks in the market.

Risk and Stock Value

PRINCIPLE 12. The effects of firm-specific risk are impounded in stock prices through cash-flow expectations, not required investor returns.

A change in stock price due to any particular circumstance does not necessarily indicate a change in relevant risk to the diversified investor.

While risk and loss of stock value may be related, they are fundamentally distinct concepts. Some perceive that a loss of value increases exposure to risk. Counterintuitively, losing value can actually reduce risk exposure.

To illustrate, a hypothetical utility begins construction on a power plant based on carbon-capture technology presumed to be cost-effective. The utility's regulator, however, makes no prior commitment to cost recovery and essentially defers the determination of prudence. Halfway through construction, the utility encounters unexpected difficulties and cost overruns, so the project is abandoned. The regulator finds that because the plant is not "used and useful" to ratepayers, cost recovery should not be granted. In other words, utility investors will be forced to bear the sunk or "stranded" costs of the project. The utility's stock price declines by 25 percent in response to the regulatory decision. No doubt, this loss of value would trigger a sense of regret among utility managers and non-diversified equity investors.

The value outcome is directly observable because the utility is less valuable after the denial of cost recovery. But when was the utility "at risk" in this scenario? Since risk is essentially the possibility of being wrong about an expected future outcome, the utility was at risk *before* the regulators denied recovery. Once regulators rejected the utility's request, the risk of non-recovery of costs became a certainty. Other things being equal, the utility is actually less risky *after* the regulatory decision.

$$\text{Price of stock} = \frac{\textbf{Expected cash flow}}{\textbf{Cost of equity}} \quad \begin{matrix} \leftarrow \text{Firm-specific and sector-specific risks affect this} \\ \leftarrow \text{Systematic risk affects this} \end{matrix}$$

FIGURE 8. How risk affects stock prices

The example can be fine-tuned to demonstrate that risk and loss of value are distinct in a prospective sense as well. Assume that before the regulators make their decision, investors start to realize that the utility might not recover a cost that they thought was a certainty. This will surely put downward pressure on the stock, but that does not mean that the stock is more risky to equity investors. Investors can diversify away firm-specific risks in a stock portfolio; if there is no change in the risks that are important to equity investors, then there is no change in the cost of equity.

In simple conceptual terms, stock prices reflect a ratio of two variables: expected cash flow divided by the cost of equity (figure 8). The expected (or forecast) cash flow is affected by firm-specific and sector-specific risk; the cost of equity is affected by systematic risk only (see Principle 11). In other words, for publicly traded companies in an efficient capital market, all relevant (systematic) risks are impounded in stock prices through the cost of equity. The stock price declines in the wake of adverse firm-specific events (including those associated with managerial and regulatory actions) when the market for equity capital reduces cash-flow expectations (the numerator) relative to a constant cost of equity (the denominator).[22]

The cash a firm generates is a function of its specific operations in specific markets. The cost of equity reflects what investors can earn by purchasing the stocks of other firms with similar risk. The stock price represents not only how well the firm performs in isolation (its cash-flow potential) but also how well it performs relative to other firms as measured by the cost of equity. A highly successful firm does not just make money; it makes more money than similarly situated companies. Still, changes in one firm's cash flows do not change the expected returns on other securities. For firm-specific risk, changes in firm value derive from changes in the cash-flow forecast, not changes in the cost of equity.

Some experiences may alter probabilities for the recurrence of the same or related experiences. However, once a unique adversity happens (and assuming the same exact matter cannot recur), the risk of it happening falls to zero. The value of the company has gone down, but so has the risk. From a prospective view, potential loss of value is a risk only if it cannot be diversified away in a portfolio (a reiterated theme here). In other words, a change in stock price due to any particular circumstance does not necessarily indicate a change in relevant risk to the diversified equity investor.

Simply because firm-specific or diversifiable risks do not affect the cost of equity does not mean these risks are or should be ignored by utility managers or regulators. In fact, from a strictly financial perspective, the job of the corporate manager is to maximize the value of the firm. It is difficult for managers to control against systematic risks, but firm-specific risks remain within their purview. Bluntly, however, managing firm-specific risk to lower the cost of equity is misplaced effort.

Prudent and skilled management of firm value means identifying not only the risks that can harm the firm but also those that can be exploited to an advantage (see Principle 4 and Damodaran 2008, 374). Successful managers in any industry see risk as opportunity and spend as much time, if not more, embracing risk as avoiding it. Risk management can enhance bond values by lowering the cost of debt, since all risks affect the cost of debt capital. Managing firm-specific risk can thus enhance stock values not by lowering the cost of equity, but by increasing expected cash flows.

Interestingly, while finance theory has much to say about risk and the cost of capital, it provides little guidance with regard to managing the risks affecting cash flow.[23] Those risks are by definition specific to the firm and do not lend themselves to generalization. Logically, all firm managers should engage in risk management, consistent with their obligation to maximize firm value.

Sensitivity to Systematic Risk

PRINCIPLE 13. The key to risk assessment for equities is measuring the sensitivity of stock returns to changes in the value of the broad equity market.

No relevant benchmark exists outside of the utilities sector for regulators to use when applying the comparable-risk standard in ratemaking.

To portfolio managers, stock prices that move in concert with, or that exaggerate, broad changes in the equity markets are regarded as more risky than those that move independently of, or mute the effects of, broad market changes. The beta coefficient provides an empirical estimate of this sensitivity.[24] Stocks that are more sensitive to changes in general market conditions as measured by beta are more risky even to diversified investors.

Different investment sectors (defined functionally) demonstrate different betas, that is, different risk profiles relative to the market. Finance theory lays the foundation for calculating beta coefficients. Stocks in the financial, industrial, and materials sectors, all of which are affected the most when macroeconomic conditions change, experience greater-than-average losses when the market tumbles and greater-than-average gains when it rises. Stocks in the utility, health-care, and consumer-staples sectors also suffer losses when the market falls and enjoy gains when the market rises, but the magnitude of those changes will be smaller than that for the riskier firms.

Stocks in the financial, industrial, and materials sectors have high betas (above 1.00); stocks in utilities, health care, and consumer staples have low betas (below 1.00). Because high-beta stocks are more risky to portfolio investors, investors require

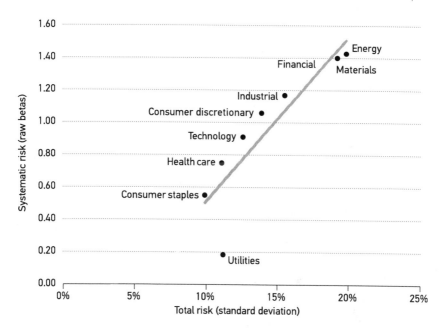

FIGURE 9. Systematic risk and total risk by sector

SOURCE: DATA FROM ALTAVISTA, *ETF ANALYZER: SELECT SECTOR SPDR EDITION,* JUNE 2014.

higher returns on those stocks; because low-beta stocks are less risky to portfolio investors, investors require lower returns on those stocks.

Risk varies within sectors. Different types of utilities (for example, electricity and water) present different risk-return profiles. Utilities tend to compensate investors with relatively steady dividend income. Generally, however, investors in the sector likely have a limited expectation for growth or appetite for risk. The utilities sector reflects low total risk and especially low relevant risk (measured by beta), even when accounting for leverage. In other words, the low-levered betas observed for utilities (reflecting both financial and business risk) suggest that they are substantially less risky than expected based on stock-price volatility (figure 9 and appendix 1).[25]

Utilities thus provide investors with proportionately greater protection from economic downturns, which clearly distinguishes the sector. In other words, not only do utilities have relatively low risk overall, represented by the standard deviation of returns, they have especially low sensitivity to systematic risk, represented by beta, which is the form of risk that actually matters to equity investors. The technical, economic, and structural traits of utilities help explain this critical distinction.

The unique risk profile of utilities is likely related to economic regulation, and this also creates a perpetual dilemma for regulators. As affirmed by the Supreme Court, and widely cited in practice, regulated utilities should be afforded a reasonable opportunity to earn returns comparable to those earned by firms facing similar risk.[26] The analysis of betas confirms that utilities are actually less risky than the stocks in all other sectors of the economy. Since there are no comparable-risk companies, no relevant benchmark exists outside of the utilities sector for regulators to use when applying the comparable-risk standard in ratemaking (see Principle 21).

Risk and Leverage

PRINCIPLE 14. Financial risk is a function of a firm's capital structure, as authorized for regulated utility companies.

When a firm is heavily leveraged, the financial markets will require higher returns on both the debt and the equity securities issued by the firm.

Companies normally finance capital projects through some combination of debt and equity, for utility companies in about equal amounts (that is, a 50–50 capital structure). For a company that relies exclusively on equity financing (no debt), the firm's earnings and return on equity (earnings divided by equity balance) vary directly with changes in economic conditions and diversifiable firm-specific risks. If instead the firm decides to use a 50–50 mix of equity and debt, fixed interest payments must be taken out of the aggregate variable earnings stream. By issuing debt, the firm has less equity to support, but the risk associated with the return on equity is higher (as measured by the range of possible returns).

Low interest rates beg the issue of substituting debt for equity. Other things being equal, however, the greater the debt, the greater the financial or credit risk associated with distress and default. This applies to households (who hold mortgages and credit cards) and corporations alike. Consider a firm that decides to deploy 90 percent debt and only 10 percent equity. Since the firm is so heavily reliant on debt, it has greater exposure to bankruptcy risk.

These mathematical relationships are illustrated in figure 10. Increasing use of debt magnifies the range of equity returns. Issuing debt acts like a lever on the firm's earnings rate, driving up high returns and driving down low returns, which is why corporate debt is often referred to as financial leverage. When a firm is heavily

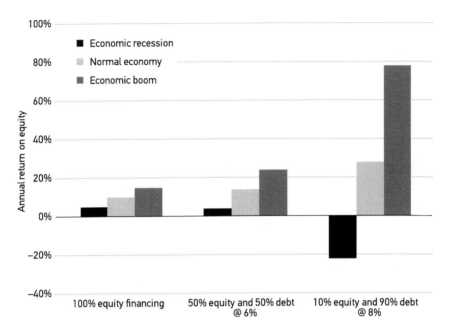

FIGURE 10. Influence of capital structure on equity returns for a single firm under alternative economic conditions (hypothetical data)

leveraged, the financial markets will require higher returns on both the debt and the equity securities issued by the firm.

Differences in leverage account for some of the differences in risk across investment sectors; for example, financial companies tend to be highly leveraged. For the utilities sector, economic regulators affect financial risk by authorizing more or less debt in a firm's capital structure (the combination of debt and equity). In other words, a utility's weighted cost of capital (debt and equity at their respective levels and rates) will reflect its financial risk (see Principle 15).

Risk and Credit Ratings

PRINCIPLE 15. A higher credit rating does not necessarily translate into a lower total cost of capital.

Good reasons might exist to secure higher bond ratings by lowering the amount of debt in the capital structure, but the prospect of lowering the overall cost of capital is not likely to be among them.

Favorable credit ratings are enticing to capital-intensive industries like utilities because they will lower the cost of debt capital and, presumptively, the total cost of capital. Strategically lowering the proportion of debt to equity will enhance credit ratings but, counterintuitively, the resulting lower cost of debt may not have the desired effect on overall capital costs.

The financial risk associated with debt is a function of the dynamic interaction of capital costs and capital structure as well as tax policy (see Brigham et al. 1987; De Fraja and Stones 2004). If a utility substitutes a substantial amount of equity for debt, four things are likely to happen: (1) the cost of debt will decline, (2) the cost of equity will decline, (3) the utility's bond rating will improve, and (4) the overall cost of capital will *increase*. Adjusting the capital structure to maintain a better rating on the firm's bonds actually makes it more, not less, expensive to raise capital, even though the cost of debt and the cost of equity both decline as financial risk is lessened.

Disregarding income taxes, the overall cost of capital is unaffected by changes in the capital structure over a broad range of debt and equity combinations (Modigliani and Miller 1958; Damodaran 2011). Table 5 provides a hypothetical example. In its initial position, a utility has a capital structure consisting of 50 percent debt (at a cost of 5.0 percent) and 50 percent equity (at a cost of 10.0 percent). The utility

TABLE 5. Illustration of a change in capital structure on the weighted cost of capital (hypothetical data)

		BEFORE EFFECT OF TAXES		AFTER EFFECT OF TAXES	
INITIAL CAPITAL STRUCTURE: **CREDIT RATING = BBB**		**PRE-TAX COST OF CAPITAL**	**PRE-TAX WEIGHTED COST OF CAPITAL**	**AFTER-TAX COST OF CAPITAL**	**AFTER-TAX WEIGHTED COST OF CAPITAL**
Debt	50%	5.0%	2.5%	3.0%	1.5%
Equity	50%	10.0%	5.0%	10.0%	5.0%
			7.5%		**6.5%**
REVISED CAPITAL STRUCTURE: **CREDIT RATING = A**		**PRE-TAX COST OF CAPITAL**	**PRE-TAX WEIGHTED COST OF CAPITAL**	**AFTER-TAX COST OF CAPITAL**	**AFTER-TAX WEIGHTED COST OF CAPITAL**
Debt	45%	4.7%	2.1%	2.8%	1.3%
Equity	55%	9.7%	5.3%	9.7%	5.3%
			7.5%		**6.6%**

Note: A 40% corporate income tax rate is assumed.

proposes to increase its equity balance to 55 percent and decrease its debt level to 45 percent. The utility's bond rating is upgraded from BBB to A, which lowers the cost of debt from 5.0 percent to 4.7 percent. The lower financial risk also lowers the cost of equity from 10.0 percent to 9.7 percent. Yet the overall pretax cost of capital remains at 7.5 percent. It is the interaction of capital cost rates and weights, not the rates alone, that determines the overall cost of capital.

However, when taxes are taken into account, substituting equity for debt will not succeed in lowering the cost of capital. Interest payments on debt instruments are tax deductible; returns to equity holders are not. Debt therefore has inherent tax advantages over equity over a range of capital structures. In other words, using equity to retire debt substitutes taxable capital for tax-deductible capital, reducing the tax shield that debt offers. As noted earlier, however, taking on too much debt increases the financial risk (the chance of bankruptcy), and the higher cost of both debt and equity eventually offset any advantage of substituting less-expensive debt for more-expensive equity.

A utility with a higher bond rating thus may have a higher overall cost of capital, even though both debt and equity costs are lower than those of a utility with a lower rating. Regulators should be wary of arguments for manipulating the capital structure in relation to bond ratings. Good reasons might exist to secure higher bond ratings by lowering the amount of debt in the capital structure, but the prospect of lowering the overall cost of capital is not likely to be among them.

The tax benefits of debt are significant over a wide range of capital structures, signaling firms to take on more debt, not less. But at some point, countervailing forces offset the benefits of debt. Having too much debt limits a firm's financing flexibility and increases the likelihood of encountering financial distress. When debt capacity is reached, firms have no choice but to turn to equity markets for financial capital, which many firms are reluctant to do because it tends to put downward pressure on stock prices. Preserving some flexibility in the capital structure allows for a nimble response to changing conditions and emerging opportunities (Damodaran 2011, 379). In other words, more equity in the capital structure provides a financing cushion, leaving debt capacity to fund projects as they come along. But maintaining that cushion does not come free. The firm with more equity enjoys a better bond rating but incurs a total cost of capital that is actually higher than for firms with less equity.

Risk and Bond Investors

PRINCIPLE 16. Bond investors cannot easily eliminate the effect of firm-specific risks through portfolio diversification.

Though relevant to bondholders, the heavy focus on firm-specific risks limits the relevance of bond-rating information to the cost of equity and equity shareholders.

Risk is experienced differently by different types of investors. Equity investors are concerned with the probability of realizing expected returns, while debt holders are concerned with the probability of credit default (Ganguin and Bilardello 2005, 80–107).

Firm-specific risks are unimportant to diversified equity investors but relevant to bondholders due to the nature of bond contracts. Simply put, bonds are debt instruments that provide a stable interest payment. While risk for the diversified equity portfolio is bidirectional (that is, with upside and downside), bondholder risk is unidirectional because bondholders do not benefit from a firm's *upside* gains.[27] This allows firm-specific risk to creep into the cost of debt, even for debt held in a diversified portfolio. As Damodaran explains (2001, 175):

> In contrast to the general risk and return models for equity, which evaluate the effects of market risk on expected returns, models of default risk measure the consequences of firm-specific default risk on promised returns. Although diversification can be used to explain why firm-specific risk will not be priced into expected returns for equities, the same rationale cannot be applied to securities that have limited upside potential and much greater downside potential from firm-specific events.

TABLE 6. Return possibilities for bonds and stocks under alternative economic conditions (hypothetical data)

	BANKRUPTCY	WORSE THAN EXPECTED CONDITIONS	EXPECTED CONDITIONS	BETTER THAN EXPECTED CONDITIONS
Bonds	−50%	5%	5%	5%
Stocks	−100%	0%	9%	14%

Bond coupon rates are fixed at the time a bond is issued and will stay fixed over a defined period regardless of whether the issuing firm is extraordinarily profitable (see Damodaran 2011, 87). In other words, corporations never pay *more* than they promise to pay to bondholders (that is, the bondholder either gets promised principal and interest payments or something less, but never more). It is no wonder that bondholders dislike risk. They might even experience a dose of regret when equity holders capture substantial upside benefits.

The relevant risk to bondholders is whether the utility will fail to deliver on promised payments, either in part or in full. As an example (table 6), if the bond has a stated interest rate of 5 percent, the most the bondholder will receive from the utility is the 5 percent interest payment. If the company that issued the bond defaults, the bondholder will get something less than 5 percent. Even if the issuing company becomes the next big success in its sector, the bondholder still earns only 5 percent.

This asymmetry explains why firm-specific risk affects the cost of debt, even though it does not affect the cost of equity. The cost of debt is based on the likelihood that the firm will default, which (absent an economic meltdown) is largely a function of firm-specific risks. The mere chance that a future default might occur can affect the value of a utility bond. If investors today believe that there is an increased chance that a utility will default on its bond obligations, then the price of the bond will decline. If they hold the bond to maturity, and if the utility ends up making all scheduled interest and principal payments, the bondholders will escape any impact. But if the investor wishes to sell the bond prior to its scheduled retirement date, the impact of risk will be manifested in bond sale proceeds that are lower than they would have been absent the change in risk.[28]

Nonetheless, it is important to avoid overstating bondholder risks. Holding both stocks and bonds and holding many bonds (or a bond fund) will mute the impact of an individual default on a bond portfolio. Finance principles hold that bonds are still less risky than stocks, as illustrated in table 6. In the vast majority of cases, bonds pay off as expected. Although they result in a substantial loss of value to bondholders, defaults are the exception (Moody's Investors Service 2014b).

When compared with equity holders, debt holders face more risk *factors* but not more risk *overall*. In an efficient market, shareholders expect to make more money than bondholders. The higher return on the stock reflects its higher level of risk. A diversified equity portfolio can be pummeled substantially by changes in macroeconomic conditions (systematic risks). But while stocks can be affected in large ways by only a few systematic factors, bonds can be affected in small ways by numerous systematic and firm-specific factors. The risk faced by a bond investor is like a series of many small pokes that generally cause little cumulative damage. The risk faced by an equity investor is more like the landing of one of a few devastating punches. In the end, higher risk is reflected in the greater dispersion of returns in stock portfolios as compared to bond portfolios.[29]

These differences are important to risk assessment. Credit reports for bonds often include a long list of risk factors, most of which are firm-specific and relevant to the firm's bondholders but irrelevant to its shareholders (because they do not drive equity portfolio values). As discussed later, it also is unsurprising that bond investors and credit analysts welcome cost-recovery and revenue-assurance mechanisms that tend to insulate individual utilities from all risk (both systematic and firm-specific) while nominal bond rates remain constant. Though relevant to bondholders, the heavy focus on firm-specific risks limits the relevance of bond-rating information to the cost of equity and equity shareholders.

Risk and Utility Managers

PRINCIPLE 17. Utility managers are sensitive to both systematic and firm-specific risks.

Given sensitivity to firm-specific risk, corporate leadership and culture have a direct bearing on how risk is perceived and managed.

U.S. corporate culture is relatively risk-sensitive, perhaps due in part to the broader financial and economic regulatory environment. The annual reports of U.S. companies with publicly traded stock (required by the Securities and Exchange Commission) typically disclose numerous downside "risks," many if not most of which are idiosyncratic (firm-specific). These risks may be relevant to bondholders and non-diversified equity investors, but are irrelevant to diversified equity investors. The corporate culture of U.S. public utilities seems particularly sensitive to issues of risk, likely due to the nature of utility services, legal considerations, and regulated returns. Risk concepts are invoked frequently, but often without precision with regard to the difference between risk and uncertainty and the reality that risk has upside as well as downside potential.

Although utility executives work for shareholders, their individual risk profiles tend to mimic those of bondholders, especially in terms of sensitivity to firm-specific risks. In reality, managers actually are more like non-diversified equity holders because they can benefit from firm-specific upside gains, which bondholders cannot.

For most intents and purposes, managers are non-diversified and cannot easily diversify their personal exposure to the risks their utility faces. Managers understandably are sensitive to firm-specific risk because they have a personal stake in the firm's failure or success. Firm-specific risks affect the manager's personal well-being, namely, their employment and compensation. Managers also tend to

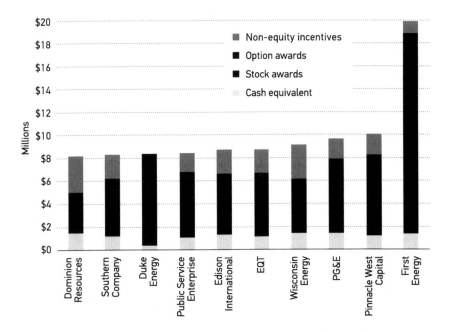

FIGURE 11. Annual compensation for the top energy utility CEOs in the United States, 2012
SOURCE: DATA FROM SNL FINANCIAL (2013).

have at least some of their personal wealth (if they own the utility's stock or are compensated with stock options) and all of their wages tied to the fortunes of the firm (see figure 11).

When a firm-specific cataclysm occurs, causing a utility's stock price to fall to zero and a default on bonds, executives who had invested heavily in the company's stock will take a hard hit, losing substantial wealth. Even worse, these managers could be fired and lose their entire income stream. Managers stand to gain when major projects succeed (raises and promotions) but can lose a lot when they fail (jobs and professional reputations). An extreme case of malfeasance might hold a manager personally liable. In relative terms, the diversified equity investor is least affected, the bond investor is somewhat affected, and the manager is most affected by a corporate failure. It should come as no surprise, then, that utility managers would seek to reduce exposure to risks perceived as mainly to the downside.

Decision making by utility managers can thus be substantially influenced by ambiguity and potential for regret. When motives and incentives conflict, principal-agent issues arise. Given sensitivity to firm-specific risk, corporate leadership and culture have a direct bearing on how risk is perceived and managed. Risk is a prime motivator for managers that cuts both ways. Astute managers see many forms of risk in terms of not just threat but opportunity. Accordingly, intelligent risk management can set utilities and their managers apart. Systems of performance review and compensation can be designed accordingly. In fact, regulators might consider whether it would prove less costly and more effective to reward managers rather than investors for risk-taking and innovation (Kihm 1991).

Regulators generally are reluctant to micromanage utility companies, but performance-based regulation suggests a more deliberate approach (see Principle 23). For reasons of logical alignment and fairness, those who will benefit from performance improvement should financially support any performance-based compensation. In other words, ratepayers should bear the cost of incentives intended to produce ratepayer value, but shareholders should bear the cost of incentives intended to produce shareholder value. If shareholders stand to benefit from incentive-based executive compensation, for example, they should pay for it. If both ratepayers and shareholders stand to benefit, then sharing the burden makes sense.

Risk and Ratepayers

PRINCIPLE 18. **Utility ratepayers typically are captive with regard to many costs and risks, and must rely on regulators to protect their interests.**

Utility managers have more capacity to identify, understand, and manage various types of risk than all but the most sophisticated ratepayers.

Public utility ratepayers are considered "captive customers" because these essential services are largely undifferentiated and choices range from limited to nonexistent (that is, purchasing is not "diversifiable").[30] In the context of monopoly, ratepayer value is defined primarily in terms of the cost of service for a particular level of service. Ratepayers are effectively "cost-takers," meaning that they are relatively powerless in the short term, at least, to affect the costs that affect them through utility prices. Similarly, ratepayers are "risk-takers" to the extent that they also bear the risks assigned to them primarily through the regulatory and deregulatory processes.

As a consequence of deregulated natural-gas commodity prices, for example, utility customers today directly experience short-term and long-term volatility in natural-gas prices (figure 12) without the intervening influence and buffering effects of either utilities or regulators.

Utility ratepayers are not unlike utility bondholders and utility managers with regard to both systematic and firm-specific risks (discussed in Principles 16 and 17). In other words, for good or for bad, consumers can be affected by both broad economic conditions and the performance of their utility service providers. In fact, due to the expansive use of various adjustment mechanisms between rate cases, ratepayers may be more likely to absorb risks associated with cost and revenue

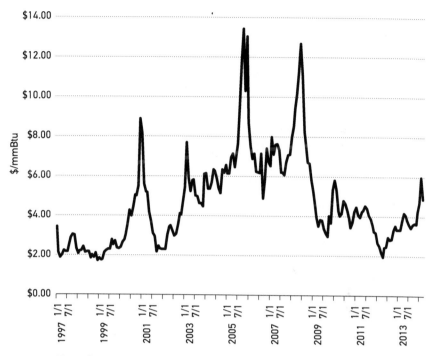

FIGURE 12. Natural-gas prices (Henry Hub Spot Price), 1997–2014

SOURCE: DATA FROM THE U.S. ENERGY INFORMATION ADMINISTRATION (WWW.EIA.GOV).

variability, whatever the reason (see Principle 27). In general, it is unclear whether ratepayers will benefit from the conditions that favor equity shareholders. For example, benefits from growth may depend on the realization of scale economies or other operational efficiencies that lower the cost of service. For this reason, regulators tend to consider ratepayer impacts in the context of capital planning and merger cases.

Of course, the economic regulator is positioned between the utility and the ratepayer, and regulatory policies and decisions (as discussed in the following principles) have a direct bearing on risk allocation. The regulator, in other words, decides whether outcomes associated with various risks will flow to utility investors or ratepayers or both. Because choices for most ratepayers are constrained, the capacity for risk management is asymmetric and favors utilities. Some large-volume utility customers (industrial, commercial, and aggregators) may be positioned to install backup systems, engage in physical or financial hedging, or implement other risk-management strategies. For the most part, however, utility managers have more capacity to identify, understand, and manage various types of risk than all but the most sophisticated ratepayers.

Regulatory Risk

PRINCIPLE 19. Economic regulation and regulatory risk substitute for competition and competitive risk.

It may be counterproductive to reduce risk exposure when beneficial investment, efficiency, and innovation by public utilities are desired.

Because traditional public utility enterprises are monopolistic, economic regulation "in the public interest" provides a proxy or substitute competition to prevent abuse of market power and its deleterious effects (such as price discrimination, service quality degradation, and supply curtailment). Regulatory accountability allows for the deployment of private capital in utility infrastructure as well as profit-driven performance incentives. In the absence of competition, regulation imposes performance discipline to encourage efficiency and innovation. Both competition and regulation should drive prices toward efficient levels. Regulation must be a fair but tough substitute for market competition because competition is tough (that is, inherently *risky*).

Regulation constitutes a historical political compromise between public ownership and competitive markets. In the early evolution of utilities, competition was perceived as risky, if not potentially "ruinous," given the unique capital requirements and cost profiles of utilities. In effect, economic regulation maintains more risk than government ownership but less risk than competitive markets (figure 13), and in some respects makes privately owned utilities an indirect arm or agent of the state.[31] Regulation thus serves not just financial but political purposes, including consideration of social goals as well as "fairness" in risk allocation between investors and ratepayers (see Myers 1972). Nonetheless, regulation was conceived to substitute

FIGURE 13. Regulation as a historical compromise and implications for risk

for competition but not for utility management (including management of certain forms of risk).

As noted earlier, regulatory risk is generally firm-specific; individual regulatory agencies and their decisions cannot affect systematic or sector risk, although they may affect a utility's sensitivity to these broader risks. To the extent that regulation changes exposure to any particular risk, the utility's bondholders, ratepayers, and managers will be affected. To the extent that regulation changes sensitivity to systematic risk, the firm's shareholders will be affected. A persistent policy question concerns the capacity of regulatory institutions to respond to changes in macroeconomic conditions in a timely and sufficient manner (Congressional Budget Office 1986).

Risk drives performance (efficiency and innovation) in both regulated and competitive firms, perhaps more so in the latter given differences in opportunities. Institutionally, economic regulation is not meant to eliminate firm-specific risk but to proportion it fairly. Risk-bearing is part of the regulatory compact and relates directly to maintaining the investment and profit motives of investor-owned utilities (see Principle 20). Regulation was never designed to provide utilities with guaranteed or risk-free returns on their investment (see Principle 22). If it were, the cost of utility debt would approach the yield on Treasury bonds. Indeed, absolving utilities from risk, including systematic risk, is suggestive of making them a direct arm of the state (that is, public ownership). Throughout its history, regulation has always dealt with the challenges of risk and uncertainty, either explicitly or implicitly. Like all firms, regulated utilities experience risk-related tradeoffs within a larger context. Regulators in turn seek to align risk and returns with the public interest while protecting captive ratepayers from the deleterious effects of excessive risk-taking. When reality does not match expectations (a potential cause of regret), regulators must exercise restraint in revisiting risk allocation after the fact. Mitigating the effect of downside risks on utilities, even in extreme cases, distorts incentives and invites moral hazard.

Utilities and their investors strongly favor regulatory stability and certainty. Obviously, regulation shuld not be arbitrary or capricious, or overtly politicized, but it must be diligent. Regulation works best when institutional roles and responsibilities are clearly delineated. A contemporary challenge for utilities is compliance with various forms of regulation (economic, financial, and environmental) that have competing performance objectives, standards, and timetables. Although differentiated and complementary regulation may be desirable (for example, environmental and economic accountability), disharmonious governmental rules can lead to suboptimal behavior and unintended consequences.

In the economic regulatory realm, unfavorable treatment of costs may increase a utility's cost of debt, but the cost of equity will increase only to the extent that a disallowance affects the utility's sensitivity to systematic risk. While certainty of cost recovery (and thus returns) might reduce the cost of capital, it might also exact a social cost in terms of efficiency losses associated with the reduction of utility risk and associated performance incentives. In the long run, regulatory certainty limits losses to investors, but it will also limit rewards. Legislative and regulatory policies that reduce risk exposure without reducing the potential for returns will likely produce inefficient and inequitable windfalls to utility investors. It may be counterproductive to reduce risk exposure when beneficial investment, efficiency, and innovation by public utilities are desired.

Risk and the Regulatory Compact

PRINCIPLE 20. Economic regulation of public utilities centers on a social compact that establishes regulatory risk and a framework for risk allocation.

Reasonably allocated risk under the regulatory compact provides public utilities a path to profitability as well as essential performance incentives.

Given the essential nature of public utilities, an accepted construct known as the social or regulatory "compact" establishes a system of risk allocation to serve the public interest in terms of both efficiency and equity. The compact serves a legislative purpose and assigns conditional rights and obligations to utilities in order to balance their interests with those of the captive ratepayers they serve (table 7).[32] Reasonably allocated risk under the regulatory compact provides public utilities a path to profitability as well as essential performance incentives. Regulation imposes discipline and holds utilities accountable, but it also affords an institutional framework for cost recovery and returns not afforded by competitive markets (see Principle 21).[33]

In theory and practice, regulatory risk suggests that regulators have considerable power and discretion to choose how much risk utilities must bear and when to compensate investors in order to "keep them whole" in the face of risk. Because regulators decide how risk is shared, both investors and ratepayers bear regulatory risk. Regulatory risk can be a salient issue for bondholders because they cannot diversify away this firm-specific factor, but it does not generally alter the systematic risk that affects equity shareholders. Changes in regulatory policies may substantially affect a utility's stock price (by affecting cash flow), but few if any regulatory decisions represent *relevant risk*, which determines the cost of equity. Recalling the

TABLE 7. Rights and obligations under the regulatory compact

UTILITIES RIGHTS	UTILITY OBLIGATIONS
The regulated utility enjoys: A conditional exclusive franchise for a certificated service territory, rights of eminent domain, protection from direct competition and antitrust, recovery of costs through rates charged, and a reasonable opportunity to earn a fair return on prudent and useful investment.	*The regulated utility accepts:* An obligation to provide all paying customers with access to safe, adequate, reliable, convenient, and nondiscriminatory service on just and reasonable terms, while assuming certain business and market risks and subjecting itself to comprehensive regulatory review and oversight.

covariance principle, it is not the individual stock price variance that matters but the covariance of stock prices in a portfolio.

This is another critical point in the context of regulation. A regulatory action that unexpectedly causes a utility stock price to decline typically does not increase the relevant risk of the utility stock (see Principle 11). For example, a lower-than-expected authorized return may affect investment value but not necessarily relevant risk. As emphasized here, firm-specific risks affect the expected cash flow of the utility, not the diversified investors' required return. The stock price declines because the utility is likely to be less profitable after the regulatory change, not because of a change in risk. In sum, the increased potential for value losses in the utility stock price (for any reason) should not be confused with an increase in risk.

Regulation generally provides a means of cost recovery for prudent and necessary or mandated investments and expenditures. But regulation should not shield utilities from business risks related to operational performance and consumer preferences, and systematic risks related to broader economic forces. Regulators, in other words, should focus on supplanting the risks that would otherwise arise from competition to drive performance. Under the compact, utility returns (profits) are *authorized* but not *guaranteed.* The authorized return sets a goal for the utility—one that often is not reached (acting more as a ceiling than a threshold). Arguably, achieving returns should be viewed as an ongoing challenge (not an entitlement), just as in a competitive environment. The same logic pertains to alternative (nonutility) service providers who, in the interest of economic efficiency, should also be subject to either competitive or regulatory risk.

Although the regulatory compact actually provides a very reasonable framework for allocating risk, regulators may be tempted to be more "proactive" and try to "reduce risk" to utilities, presumably to lower costs (namely, the cost of capital) to consumers. Such good intentions might be misplaced. For example, regulatory policies that essentially provide "preapproval" of cost recovery depart from, if not void, the compact and associated incentives by negating the chance of unrecovered costs or stranded investment.[34] Raising rates and authorized returns to utilities in the face of risk does not alter their risk profile. In fact, a preponderance of economic theory clearly suggests that competition and competitive risk should lead not to higher but to lower prices and profit margins (Porter 1998, 5).

Managing utility exposure to risk is not a primary regulatory objective, and the regulatory process is not structured for this purpose. Moreover, though perhaps counterintuitive, risk reduction may not be desirable and might even be costly, particularly in terms of economic efficiency. Properly allocated risk under the regulatory compact provides utilities with essential performance incentives. Privatized risk is managed by markets, which are presumed to be efficient in pricing risk. Socialized risk is *assumed* by the government but *borne* by taxpayers or ratepayers, all but eliminating market forces.[35] Assuming risk undermines regulation as an institution because the regulator's role shifts from arbiter to administrator.

Risk and Regulatory Standards

PRINCIPLE 21. **Accepted standards of regulatory review do not insulate utilities from risk or guarantee returns on investment.**

Regulatory risk, though substantial, is a bounded form of risk compared to the risk faced by competitive firms largely because of regulatory jurisprudence.

Economic regulation of public utilities in the United States follows a series of recognized and synergistic standards of review applied by commissions and courts alike. In many respects, the history of U.S. regulation can be told through a history of Supreme Court precedents, many of which speak directly to the issue of risk. Utilities and other stakeholders have rights of appeal, of course, if they believe a standard or their constitutional rights have been violated.

Several of the core risk-related standards of review applied in regulation are summarized in table 8 (and detailed in appendix 2). These center on a constitutional requirement to properly compensate private investors for use of their property. Although constitutionally grounded, regulatory standards allow for judgment and pragmatism. Still, regulators across federal and state jurisdictions follow relatively consistent approaches that establish relatively consistent expectations. By operating according to widely understood and accepted standards, regulation exerts a stabilizing force. The prudence test (see Principle 23), for example, can limit risk exposure for both utilities and ratepayers. The economic traits of utilities along with the standards and practices of regulation help explain why the utilities sector has a distinctly low-risk profile. Thus, regulatory risk, though substantial, is a bounded form of risk compared to the risk faced by competitive firms largely because of regulatory jurisprudence.

TABLE 8. Standards of review applied in regulation

- Returns are authorized but not guaranteed and they cannot place "unjust burdens" on ratepayers (*Smyth v. Ames*, 1898).
- Regulation involves the "fair interpretation of a bargain" that finds a "midway" between too little and too much profit (*Cedar Rapids*, 1912).
- Returns should reflect "corresponding risk" to maintain credit and attract capital while balancing "wealth and welfare" (*Bluefield Water Works v. WV PSC*, 1923; *FPC v. Hope Natural Gas*, 1944).
- Commission judgment should not substitute for board discretion, and prudence is presumed, but "dishonest, wasteful, or imprudent expenditures" should be disallowed (*Southwestern Bell v. Mo. PSC*, 1923).
- Utilities are not entitled to include in rate base property that is not "used and useful" to ratepayers (*Denver Union Stock Yard*, 1938).
- Regulation does not ensure that businesses will produce "net revenues" or recover losses (*FPC v. Nat. Gas Pipeline*, 1942).
- Regulators are not bound by formulas and are free within their statutory authority to make "pragmatic" adjustments to achieve results (*FPC v. Natural Gas*, 1942; *FPC v. Hope*, 1944).
- Due process does not insure or protect utilities from losses due to business risk associated with "economic forces" (*Market St. Railway*, 1945).
- A contract rate may be evaluated relative to the public interest but is not "'unjust' or 'unreasonable' simply because it is unprofitable" (*FPC v. Sierra Pacific*, 1956).
- Utilities must "operate with all reasonable economies" (*El Paso Nat. Gas*, 1960).
- Regulators should not usurp management or judge prudence with the "20-20 vision of hindsight" (*TWA v. CAB*, 1967).
- Courts allow regulators to decide within a "zone of reasonableness" (*Permian Basin Area Rate Cases*, 1968).
- Rates are "prospective in nature," and "the utility company must bear the risk of loss inherent in the well-known lag" in ratemaking (*Narragansett*, 1977).
- Utility monopolies are "relatively immune to the usual market risks," so risk is largely defined by rate methodologies that should not arbitrarily shift risks to and from investors (*Duquesne Light v. Barasch*, 1989).

Although regulation takes on a judicial form, it is also recognized as a legislative function. Regulatory commissions generally have considerable discretion, which translates to regulatory risk, but their discretion also is bounded by legislative and judicial policy. Legislatures are free, for example, to change the rules that define recoverable costs and cost-recovery methods. But while utilities get a degree of financial cover from regulation, at least in the short run, they are not exempt from long-term technological, economic, and structural dynamics. Examples can be found in the transformation of telecommunications services, and more recently in the expansion of distributed energy resources (see Principles 30 and 31).

Risk and Fair Returns

PRINCIPLE 22. Fair returns are set to exceed the utility's cost of capital not to compensate for risk but to encourage socially beneficial investment.

A return premium serves a social purpose because a compensatory return equal only to the cost of capital makes utilities indifferent to public policy goals attached to their performance.

In many respects, the central job of the economic regulator is to divine the "fair-return" price for utility monopolies. Left to their own devices, unregulated monopolists will tend to withhold supplies and escalate prices, resulting in a dead-weight loss to society. As noted previously, effective competition in any sector will drive prices to socially optimal marginal costs of production.[36] Economic profit is achieved only by finding a competitive advantage, discovered through efficiency or innovation. When prices are driven to cost, competitors turn to non-price considerations, namely, unique product or service offerings (see Porter 1998).

To promote economic efficiency by replicating competitive forces, regulators might be tempted to set prices equal to marginal costs (also understood as socially optimal prices). This can be a problem for traditional capital-intensive utility monopolies with pronounced scale economies because the marginal cost of production tends to fall below the average cost of production, which includes capital-related costs. Prices at marginal cost will not adequately compensate the utility or ensure that it will make appropriate investments to meet service obligations or achieve social goals. Fair-return prices for utility services will yield returns that fall somewhere between constitutionally confiscatory and excessive relative to what the market demands in terms of comparable risk because they serve policy purposes (particularly adequate supply of essential services). Because their private property is devoted to public service, the utilities are entitled to just compensation under the "takings" clause of the U.S. Constitution,[37] including a return of (depreciation expense) and on (return on equity) investment.

A rate that is merely compensatory (at the opportunity cost of capital) will be sufficient to attract capital relative to risk, but it will neither motivate nor reward performance. Judicial precedents emphasize that the "right" price and profit balances the interest of utilities ("wealth") and their ratepayers ("welfare").[38] The fair-return price will not only compensate the utility for the cost of capital but will also attract capital based on earnings (versus risk) comparability. The Supreme Court allows considerable discretion to regulators with regard to choice of methodology as well as the application of "pragmatic" adjustments (within statutory boundaries) to achieve desired goals (that is, end results matter more than means). In the end, regulated prices and returns are expected to be "just and reasonable."

Finance theory holds that if investors expect a firm to earn only its cost of capital over the long run, the firm's stock price will equal its current book value (Fairfield 1994; Kolbe, Read, and Hall 1984).[39] Put another way, stock prices remain constant when the rate of return equals the risk-adjusted cost of capital. For U.S. utilities, market-to-book ratios exceed a value of 1.0 by a considerable margin, due primarily to regulatory policy with regard to allowed returns. Applying the commonly used discounted cash-flow model, a utility with a 4 percent dividend yield and 4 percent expected long-run dividend growth, for example, would have a cost of equity equaling 8 percent.

As summarized in table 9, the cost of capital and the authorized return form a band of acceptance. The *cost* of equity demarks a lower bound and a starting point for establishing the *fair return* on equity (see Kahn 1988; Bonbright et al. 1988; and Phillips 1993). Setting the return above the cost of capital is essentially a judgment call that is informed by financial and economic analysis, but also by legal and policy considerations (Kihm 2007; see also Morin 2006). While much is made of the need to avoid a confiscatory return (too low), the Supreme Court also limits returns to the upside (too high).[40] Excessive earnings suggest the need for regulatory intervention in terms of revisiting authorized returns.

The difference between the risk-determined cost of equity (for comparable risk) and the regulatory-determined fair return (for commensurate or comparable earnings) can be accurately defined as a sufficient *return premium* (that is, not a "risk premium"). A fair return compensates the utility for prudent performance

TABLE 9. Regulatory consideration of risk and policy in setting equity rates of return

AUTHORIZATION	RETURN	EXPLANATION
Regulatory consideration of policy	∧ Excessive or extortive return	An economically inefficient return
	∧ Incentive or bonus return	A return with a premium to motivate desired performance
	∧ Fair return	A return with a premium to motivate beneficial investment
Regulatory consideration of risk	∧ Compensatory return	A return based on the cost of equity including an equity-risk premium
	∧ Risk-free return	A return based on the yield on risk-free securities*
	∧ Confiscatory return	A return below the cost of equity (unconstitutional taking)**

* Government-owned and not-for-profit utilities are generally insulated from equity risk.
** For an investor-owned utility that still faces equity risk, any return below the cost of equity would be considered confiscatory.

(substituting for competition) and encourages efficiency (Breyer 1982, 47). Sufficient return opportunities also promote infrastructure investment, productivity improvement, and economic progress, as articulated by Kahn (1988, 44):

> The cost of capital is only the beginning point. . . . [It] does not necessarily tell us how best to promote economic progress. The provision of incentives and the wherewithal for dynamic improvements in efficiency and innovations in service may require allowing returns to exceed that level. . . . Thus, the rate of return must fulfill what we may term an institutional function: it somehow must provide the incentives to private management that competition and profit-maximization are supposed to provide in the nonregulated private economy generally.

Kahn (1988) recognized that fair returns provide incentives that compensatory returns do not. Returns based on risk alone, in other words, reinforce a neutral and static view of needs and preferences. A return premium serves a social purpose because a compensatory return equal only to the cost of capital makes utilities indifferent to public policy goals attached to their performance. In fact, although compensatory, setting the allowed rate of return at the cost of capital sends a signal to the utility to be passive in all aspects of its business operations.[41] In financial terms, the utility would do just as well to increase output, decrease output, or simply close down operations and sell its assets, assuming it were unencumbered by an obligation to serve (Train 1991). Evidence suggests that entry barriers and other market imperfections enable many firms (regulated and competitive) to realize returns above the cost capital (Damodaran 2011, 318; Brilliant and Collins 2014), which validates the consideration of comparable earnings when determining return premia.

Importantly, when setting the overall rate of return, the regulatory consideration of risk is effectively limited because relevant risk is already factored into the cost of debt and cost of equity to arrive at a *compensatory* return (the weighted cost of capital). The cost of debt is revealed by the bond market; the cost of equity is revealed by a comparable-risk (or cost-of-capital) analysis using generally accepted methodologies and considerations, including capital-project scale. The risk-based cost of equity must be correctly determined before establishing any policy-based equity return premium. Regulators should also take into account cost-recovery and revenue-assurance mechanisms when analyzing comparable risk to determine the cost of equity (see Principle 27).

The cost of equity is an estimate of investor expectations based on financial analysis; the return on equity capital can be observed after the fact based on accounting data. If the return on equity exceeds the cost of equity, the stock price will exceed the book value of equity; if the return on equity is lower than the cost of equity, the stock price will be lower than the book value of equity.

Firm-specific risk factors affect the total cost of capital because the cost of debt affects the total cost of capital. But diversifiable, firm-specific risk is not appropriately considered by regulators when determining the cost of equity. While greater risk may increase the cost of capital, it is not necessarily grounds for a higher return on equity, because the cost of equity may be unaffected; likewise, lower risk is not necessarily grounds for a lower return on equity. Again, a risk is relevant only if it exposes the utility to more or less systematic macroeconomic risk. For these reasons, regulators should be circumspect about risk information provided from a debt perspective when considering the cost of equity.

Risk and Regulatory Incentives

PRINCIPLE 23. All regulation is incentive regulation and all incentive regulation is based on regulatory risk.

The job of the regulator is not to micromanage utilities but rather to frame the system of performance goals and incentives within which utilities must manage themselves.

A widely shared principle in the field, attributed to scholar and regulator Alfred Kahn, is that *all regulation* is incentive regulation, based on the premise that public utilities will respond to the performance incentives that regulators provide. Although methods and practices vary, well-implemented economic regulation can provide powerful risk-based incentives to guard against abuse of market power as well as to promote desirable performance. Economic regulatory incentives work in concert with incentives imposed by other public policies, utility boards of directors, and broader economic forces.

As discussed under Principle 22, authorizing a return premium (that is, a return on equity above the cost of equity) can encourage socially beneficial investment, including infrastructure expansion, replacement, and modernization, and thus promote progress. The return premium is not guaranteed; it provides utilities with headroom opportunity for returns, both hitting and missing. For regulated utilities, earned returns may be greater than the cost of capital (and thus both compensatory and fair) but still less than authorized levels. Public utility managers are thus motivated to engage in return-enhancing activities.

When authorized returns exceed the cost of capital, rational utility managers will favor capital investments over operating expenditures, a propensity known as the Averch-Johnson or A-J effect (Averch and Johnson 1962). The predilection for rate-base investment generally is simultaneously regarded as a positive and negative feature of the U.S. regulatory model. In fact, regulators set rates of return that exceed the cost of capital, as seen in both utility stock prices (that is, utility stocks trade noticeably above their underlying book values) and plant-investment behavior. In the decades following World War II, utilities aggressively raised capital and expanded infrastructure networks and capacity to meet growing demand (Kahn 1988).

Regulatory oversight and standards of review (including the "prudence" and "used-and-useful" standards) are meant to guard against excessive capital investment, including "gold-plating," to ensure that the investment will be appropriate and socially beneficial, and that rates and returns will not be unduly burdensome to ratepayers.

Under the prevailing rate-base with rate-of-return regulatory model, returns are earned on the utility's rate base (approved plant in service less accumulated depreciation and net of accounting adjustments). Utility managers generally have strong incentives to invest in the rate base, seek favorable authorized returns, accelerate depreciation, recover all operating costs, pass through expenses, and meet or exceed projected sales. Utilities also are motivated to reduce risk exposure without reducing earnings potential and may pursue modifications to the regulatory model, including various ratemaking methods that shift risk from investors to ratepayers (Principle 27).

The job of the regulator is not to micromanage utilities but rather to frame the system of performance goals and incentives within which utilities must manage themselves. Information asymmetry favors utilities over consumers and other stakeholders. The use of incentives presumes that motivated utility managers will find opportunities to improve performance better than even the most astute and well-intentioned regulator. To be effective, incentives must be unambiguous and harmonious in tying rewards to performance. The effectiveness of incentives also depends on circumstances. For example, regulatory incentives tend to be stronger when costs are rising and weaker when costs are stable or falling (see Norton 1985).

Three essential risk-based incentive tools are applied in economic regulation to ensure that utility performance is consistent with the public interest (figure 14; Beecher 2013). Each is explored further in subsequent principles. Regulatory lag (Principle 24) is a passive incentive, inherent to the regulatory process, that primarily encourages cost control by the utility between rate cases. Prudence reviews (Principle 25) are reactive means of encouraging efficiency, providing essential checks on utility investments and expenditures. Finally, regulators can take a more active approach by providing incentive returns (Principle 26) that aim to encourage innovation. As they constitute returns exceeding "fair returns," these incentives should be used intelligently, that is, strategically and sparingly and with sufficient performance evaluation.

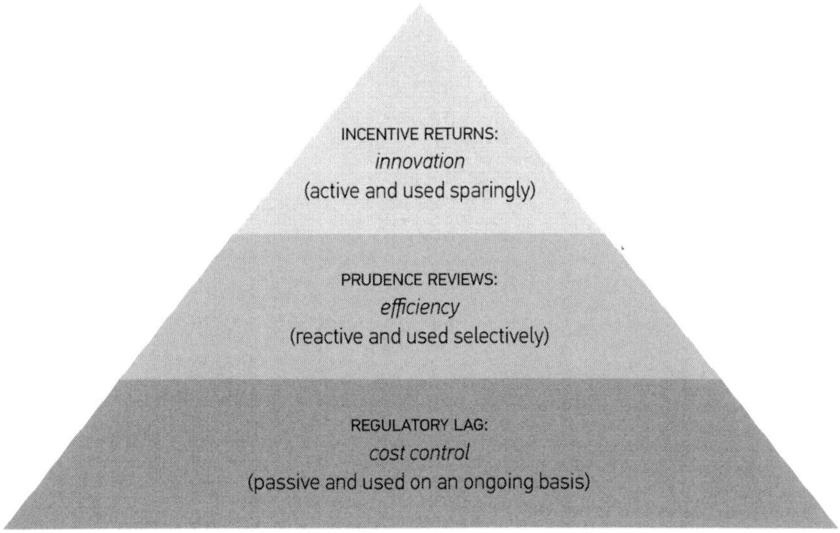

FIGURE 14. Regulatory tools for motivating public utility performance

SOURCE: ADAPTED FROM BEECHER (2013).

Risk and Regulatory Lag

PRINCIPLE 24. Regulatory lag embeds risk in the regulatory process by design and offers both upside and downside potential.

> Regulatory lag should be remediated only to the extent that it substantially jeopardizes a utility's reasonable opportunity to earn a fair return.

Although much maligned by utilities and a rather blunt policy instrument, a more balanced and nuanced view sees regulatory lag not as a bureaucratic nuisance or failure but as an incentive mechanism consistent with the regulatory compact (see Myers 1972 and Shepherd 1992). Regulatory lag is a deliberate and purposive form of risk that promotes cost control by regulated utilities. Lag combines with more active incentives imposed by utility regulators and boards of directors to motivate performance. As a matter of policy, because it promotes efficiency, regulatory lag should be remediated only to the extent that it substantially jeopardizes a utility's reasonable opportunity to earn a fair return, and the utility should bear the burden to demonstrate that this is the case.

Put simply, regulatory lag is regulatory risk by deliberate design. Formally, lag is the delay between a change in costs or revenues (+/−) and a change in authorized prices. Lag can also be understood as the time period between when an unregulated firm and a regulated firm could put in place a defensive price adjustment. Regulatory lag has various other connotations as well, some associated with different disciplines.

Cost and revenue trends following the test year used in setting prices accentuate the effect of lag, positively and negatively. By setting prices for longer periods of time, price-cap regulation formalizes lag (see Joskow 2008) and thus shifts more risk

TABLE 10. Implications of regulatory lag for realized returns based on cost, sales, and efficiency trends

		EFFICIENCY TREND BETWEEN RATE ADJUSTMENTS	
		INCREASING OPERATIONAL EFFICIENCY	DECREASING OPERATIONAL EFFICIENCY
COST AND SALES TREND BETWEEN RATE ADJUSTMENTS	FALLING COSTS AND/ OR RISING SALES	Realizing returns is likely	Realizing returns is indeterminate
	RISING COSTS AND/OR FALLING SALES	Realizing returns is indeterminate	Realizing returns is unlikely

associated with operational efficiency to investors. While not all lag is considered "good" or acceptable in practice, all lag will maintain performance pressure for cost control between rate cases. Importantly, pricing lag is not exclusive to regulated utilities. For various reasons, including competition itself, competitive firms might also be constrained in making price adjustments (see Wein 1968).

Lag in regulatory ratemaking is a function of the cumulative effects of the test year used to establish revenue requirements, timing of filings (including overlapping or "pancaking"), suspension period and statutory deadlines, complexity of issues involved, quality of the filing and evidence proffered, stipulations and settlements, and regulatory agency resources. Economic conditions, such as economic growth or decline, will tend to mask or magnify the effects of lag.

In terms of risk, although infrequently acknowledged, lag presents utilities with both upside and downside potential (table 10). As tariffed rates are set prospectively,[42] lag will work to the advantage of utilities in the context of declining costs, but to their disadvantage in the context of rising costs. In either case, efficiency and innovation following a rate case will benefit the utility. Although utility managers might complain that regulators "claw back" any gains at the subsequent rate-case opportunity, competition exerts comparable pressure on unregulated firms to force a subsequent round of effort.

Consideration of lag tends to be asymmetrical, however, because regulation is largely a reactive process, and utilities are more likely to bring attention to underearning than overearning. Consumers and other stakeholders are disadvantaged in terms of bringing attention to overearning. In other words, regulatory lag can actually work to the advantage of utilities. Moreover, not all lag is regulatory lag.

When utilities are materially affected by changes in economic or other conditions, they must be diligent and responsive, and seek appropriate regulatory relief in a timely and convincing manner. Thus, firm-specific "utility lag" may be as relevant a risk as regulatory lag.

Risk and Prudence Reviews

PRINCIPLE 25. Prudence reviews maintain regulatory risk with regard to utility investments and expenditures.

Prudent behavior is expected and earns a fair return, including a reasonable but not an extraordinary return premium.

"The practical purpose of income," according to John Hicks (1946, 172), "is to serve as a guide for prudent conduct."[43] If competition drives efficient performance in the pursuit of income, so too should regulation. Prudent performance is expected of both competitive and regulated firms. In regulation, the concept of prudence relates directly to risk. Prudence frequently is judged in risk-management terms, as the prudence test centers on whether utility managers make good decisions based on what is "known and knowable" at the time, that is, with due diligence. Drawing on regulatory case law, prudence cannot be judged with the benefit of 20-20 hindsight and is largely considered a rebuttable presumption.[44] Nor should the evaluation of prudence be clouded by a sense of regret over lost opportunities. Nonetheless, prudent performance is a condition of the monopoly's exclusive franchise under the regulatory compact.

Prudence is an especially pertinent regulatory standard with regard to risk and risk allocation. Prudence calls for anticipating and managing risk with regard to investments and expenditures. Under "normal" conditions, regulators allow for recovery of prudently incurred costs associated with investments deemed used-and-useful to ratepayers, along with a reasonable opportunity to earn a fair return on invested capital. Once prudence is well established, a considerable degree of regulatory risk dissipates. To some degree, the prudence standard thus protects ratepayers from the effects of firm-specific risks that can (and arguably should)

TABLE 11. Regulatory disallowance of nuclear costs ($ millions)

Palo Verde	$ 60.0	Callaway 1	440.0
Byron 1	101.5	Hope Creek	489.2
Wolf Creek	257.3	Perry 1	628.0
Waterford 3	284.0	Vogtle 1	951.0
Shearon Harris	322.8	Shoreham	1,395.0
San Onofre 2	344.6	River Bend	1,400.0
Limerick 1	368.9	Nine Mile Point 2	1,803.0
Fermi 2	397.0		

Source: Data from Canterbery, Johnson, and Reading (1996), 559.

be managed, while protecting investors from the effects of broad systematic risks that are essentially unmanageable. Regulation cannot and should not protect utilities from risk in all cases or at all times, as risk drives performance (as it does for competitive firms).

Prudence is effectively reviewed in every regulatory proceeding, including but not limited to those concerning certificates of need, capital improvement plans, capital structures, and methods of cost recovery. Rate cases involve regulatory scrutiny of all investment and expenditure decisions resulting in the utility's revenue requirements that will be recoverable through rates. Financial and management audits provide additional detail. A prudence review, which may be targeted or broad in scope, is used to examine managerial performance in relation to a particular investment decision or operational matter. Reviewing prudence ensures efficient use of resources by guarding against cost inflation or excess (in concert with the "used-and-useful" standard). Prudence reviews of nuclear plants resulted in substantial cost disallowances by regulators (table 11).

When regulators identify imprudence, they generally do not compel the utility to undo the imprudent deed. Rather, they impute the more efficient cost when authorizing revenue requirements and rates. For example, should a utility buy insurance at above-market rates without justification, the regulator does not ask the utility to demand a refund or cancel the policy. Instead, the difference in cost is disallowed from inclusion in rates. The disallowance comes out of the pocket of the utility shareholder, thus reducing effective earned returns. Large disallowances may draw the ire of the utility's board. Thus the possibility of disallowance can be a strong performance incentive (see Joskow and Schmalensee, 1986).

The conception of a prudent utility can and should evolve with time. Performance standards for service quality may play an increasingly important role in helping regulators gauge prudence. Regulators can judge prudence according to expectations based on contemporary performance criteria (including risk management). In sum, however, prudent behavior is expected and earns a fair return, including a reasonable but not an extraordinary return premium. Extraordinary incentives should be reserved for extraordinary performance under a considered performance-based regulatory framework.

Risk and Incentive Returns

PRINCIPLE 26. When they are deployed, incentive returns should maintain risks to motivate and reward desirable utility performance.

Extraordinary return premia should be used only to promote the achievement of specific and measurable performance goals or targets.

The conception of economic regulation as a substitute for competition relates directly to risk (see Morin 2006). Again, as discussed under Principle 22, regulated returns fulfill an "institutional function" in terms of replicating the performance discipline and incentives (both positive and negative) imposed by competition in unregulated sectors of the economy (Kahn 1988, 44). Again, however, regulated returns are authorized but not guaranteed. In other words, regulated firms are not unlike competitive firms in terms of the connection between risk and reward (or performance and profit). Regulated utilities must "reach" for returns; returns are not simply handed to them. From a technical standpoint, as also discussed, individual utilities are rewarded when their earned return (*ex post*) exceeds the expected cost of capital (*ex ante*). The degree to which the return on equity should exceed the cost of equity is a critical regulatory determination.

The traditional regulatory model generally provides more guidance with regard to reducing effective returns in the form of disallowing imprudent costs (Principle 25). Recent years, however, have seen considerable interest in providing utilities with positive incentives in the form of alternative ratemaking methods as well as "bonus" returns for performance that in some way exceeds the traditional prudence standard. Federal regulators have provided for a number of incentives for investment in high-voltage transmission (table 12), for example; state regulators have

TABLE 12. FERC-identified incentives for transmission investment

- Incentive rates of return on equity for new investment by public utilities (both traditional utilities and stand-alone transmission companies, or transcos)
- Full recovery of prudently incurred construction work in progress
- Full recovery of prudently incurred pre-operations costs
- Full recovery of prudently incurred costs of abandoned facilities
- Use of hypothetical capital structures
- Accumulated deferred income taxes for transcos
- Adjustments to book value for transco sales/purchases
- Accelerated depreciation
- Deferred cost recovery for utilities with retail rate freezes
- A higher rate of return on equity for utilities that join and/or continue to be members of transmission organizations, such as (but not limited to) regional transmission organizations and independent system operators

Source: U.S. Federal Energy Regulatory Commission summary of FERC Order 679, Promoting Transmission Investment through Pricing Reform (Docket No. RM06-4-000, Issued July 20, 2006) available at www.ferc.gov.

provided various incentives as well.[45] The use of incentives and subsidies benefits from a clear conception of market failure as well as the cost-effectiveness of these policy tools. The effectiveness of return-based regulatory incentives is not entirely clear (Lyon 2007). Rewards become excessive and inefficient if they provide too much profit relative to economic and social benefits.

The premium already embedded in the fair return serves policy purposes by compensating utilities for prudent performance *and* providing a positive performance incentive. Incentive or bonus returns go further to emulate competitive market forces that accentuate the risk-reward relationship. Incentive returns also have different consequences for different types of investors. To the extent the utility is exposed to more risk, bondholders may see the value of their bonds decline. When risky endeavors pay off, the extra returns flow only to equity shareholders.

Incentive returns attempt to exploit profit motives in the interest of promoting desirable behavior on the part of privately owned utilities. If regulators want utilities and their managers to innovate and take chances on new, potentially cost-saving technologies, for example, then incentives along these lines may be needed. But utilities should not be completely insulated from either the upside or the downside of risky activities. If utilities are given an incentive to take risks, they should truly face risk. In other words, they should reap financial rewards when things go well but suffer financial losses when things go poorly. Under some circumstances, an earnings-sharing mechanism that apportions risks and rewards between ratepayers and investors may serve policy purposes as well.

While regulatory incentives work reasonably well in terms of cost control and efficiency, innovation is more challenging. In general, regulation and innovation are not necessarily incompatible (Porter and van der Linde 1995). Providing explicit incentives for innovation, however, presumes that innovation can be readily known. Moreover, another widely held tenet is that regulators should not pick and choose particular technologies, but rather should motivate utility managers to make sound technological choices. Regulators in the UK recently initiated modifications to the price-cap regime explicitly to encourage innovation on the part of regulated utilities.[46]

Special incentives provided by regulators are sometimes rationalized on the basis that certain capital investments would be too "risky" without them, but this is better understood as a return premium to energize management toward the desired investment. In reality, utilities generally have little trouble raising capital.[47] Contrary to common belief, a higher incentive return does not attract new capital or accelerate capital attraction, but rather, it enriches existing investors (including stock-owning managers) because the effect is immediately impounded in an inflated stock price (see Myers 1972). New investors are the source of the windfall; existing investors are the beneficiaries. A key point is that utility managers work for existing (not potential) investors, who stand to benefit when managers can issue stock at higher prices. Somewhat ironically, incentives work by motivating managers to *seek* capital rather than by motivating investors to *provide* capital.

When special regulatory incentives are deployed, the regulator must determine not only the level of the incentive return but also, as importantly, the level of expected performance. Extraordinary return premia should be used only to promote the achievement of specific and measurable performance goals or targets; clear standards and metrics are needed for these purposes. Moreover, regulators must be prepared to impose consequences when goals are unmet. Failing to do so simply undermines the critical connection between risk and reward.

Risk and Ratemaking Methods

PRINCIPLE 27. Cost-recovery and revenue-assurance mechanisms shift risk between utility investors and utility ratepayers and thus affect the utility's overall cost of capital.

Certain and expedient cost recovery and revenue assurances tend to favor investors over ratepayers and should be considered when authorizing rates of return.

Once rates are set, based on revenue requirements for a test year, utilities are subject to cost and revenue risks. Over the past decades, regulated public utilities have sought a wide range of methods (including adjustments, trackers, and surcharges) to correct what they perceive as the burdensome problems of "regulatory lag" and rate-case expense (discussed under Principle 24), as well as enhance financial "resilience." These methods aim to modify the ratemaking process by expediting cost recovery and assuring revenue streams; their use will also tend to mask rate increases and mitigate rate shock. Purchased gas and fuel-cost adjustment clauses were among the first to arrive (see Clarke 1980), and the list of ratemaking modifications has evolved to include:

- Purchased natural gas adjustments
- Electricity fuel-cost adjustments
- Purchased power adjustments
- Normalization and stabilization
- Single-issue ratemaking
- Interim rates
- Cost deferrals

- Allowance for funds used during construction (AFUDC)
- Construction work in progress (CWIP) in rate base
- Attrition allowances
- Inflation adjustments
- Forward-looking test year
- Operating-cost trackers

- Accelerated depreciation
- Cost-of-service indexing
- Minimum bills
- Higher fixed charges
- Demand-repression adjustments
- Lost-revenue adjustments
- Revenue decoupling
- System-improvement surcharges
- Capital expenditure surcharges
- Securitization of stranded costs

- Project preapproval
- Rate-case time limits
- Self-implementing rates
- Cost-of-capital adjustments
- Earnings adjustments
- Higher fixed charges
- Demand charges
- Customer prepayment
- Multi-year rate plans
- Formula-rate plans

In recent years, some legislatures and regulatory commissions have been rather expansive in providing alternative ratemaking methods, and utilities may have a growing sense of entitlement to them. They are often used in combination to similar effect of reducing regulatory risk overall. Some older methods have been applied to new purposes. For example, cost-recovery mechanisms that were developed to deal with substantial, recurring, volatile, and uncontrollable operating expenses (such as fuel related) have been extended to cover certain capital expenditures (such as infrastructure improvement related) that do not clearly possess these attributes. Methods that accelerate capital-cost recovery may reduce utility exposure to some forms of risk (such as construction-cost risk) but increase ratepayer exposure to other forms of risk (such as technology-obsolescence risk).

Alternative ratemaking methods are sometimes characterized as adaptive "best practices." In addition to utilities, industry associations and other stakeholders have advocated their adoption (Lowry et al. 2013; Wharton et al. 2013). From the utility's perspective the term is apt, but what is best for the utility company and its investors may not be best for ratepayers. Furthermore, what is best today might not be best tomorrow. Some of these methods also place considerable burdens on the state and rate-case intervenors with regard to assessing prudence and applying other traditional standards of review.

While regulatory lag *maintains* regulatory risk, many if not most alternative ratemaking methods do not *reduce risk* but rather *shift risk* from utility investors to ratepayers. Risk-shifting is relevant because within their sphere of expertise, utility managers have more capacity for risk management than most of their ratepayers (Principle 18). Moreover, the ability to shift risk to others in general has the paradoxical effect of inviting more, rather than less, risky behavior.

The use of these methods raises a number of regulatory policy issues not the least of which concern the presumption of prudence and the guarantee of returns, both of which violate the terms of the regulatory compact. Expedient cost recovery is especially problematic for regulators in light of limitations on both retroactive ratemaking (that is, revisiting costs once approved) and the scope of review in reconciliation proceedings (as compared to full rate cases). The application of these methods can be less transparent as well as asymmetrical if cost increases are accounted for but concurrent cost reductions are not. Furthermore, "resolving" risk in one performance area may foreclose pursuit of efficiency opportunities in other areas. A key concern, of course, is that expedient cost recovery will dampen performance incentives by mitigating downside risk, creating a moral hazard in that utility managers will become complacent about cost control, efficiency, and prudence, and less motivated toward innovation that would enhance ratepayer value.

Turning to a revenue-assurance method, "decoupling" utility revenues from sales has been actively advanced as a means of neutralizing the perceived bias of utilities against end-use efficiency.[48] Although decoupling has diffused across several states, a risk perspective reveals that it is actually a relatively weak regulatory policy tool. Efficiency advocates tend to focus on profits alone as the key driver of utility value, even though the cost of equity and investment scale are equally important (see Principles 8 and 15). To the extent that decoupling insulates the utility from firm-specific risks (such as weather), the cost of equity will be unaltered; to the extent that decoupling insulates the utility from exposure to systematic risk (such as recession), the cost of equity will be lowered.

If an authorized return is unchanged, a decision allowing decoupling will increase the value of a utility's stock, which works to the advantage of investors but does little to alter investment incentives. In other words, as long as the return on equity exceeds the cost of equity, decoupling will motivate capital investment as a means of enhancing shareholder value. As shown in figure 15, decoupling argues for lowering both the risk premium to account for the lower cost of capital as well as the return premium to discline capital spending. In fact, decoupling would be much more effective in addressing the incentive problem if the return on equity equaled the cost of equity, but this would run counter to other goals of fair-return regulation (see Principle 22 and Kihm 2009). Decoupling does not negate other regulatory incentives for cost control and efficiency, but reducing exposure to certain firm-specific risks (such as sales) could distort performance in those areas (such as sales forecasting).

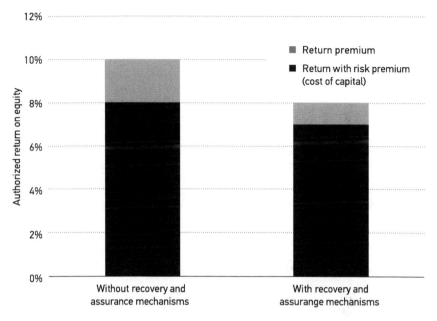

FIGURE 15. Adjustments to risk and return premia with revenue-assurance mechanisms (hypothetical data)

In sum, methods that shift risk may have unexpected or unintended consequences (see Principle 29). Moreover, heavy reliance on cost-recovery and revenue-assurance mechanisms may seriously undermine the regulatory compact and the risk-allocation framework it establishes. Regulators clearly have discretion when it comes to the use of alternative methods for addressing utility costs or revenues. However, in many respects, it might be desirable to stick with performance incentives that have been proven to work under the compact for so long. Because of implications of alternative ratemaking methods for risk, it is advisable to condition and monitor their use and to maintain regulatory risk by other means. Generally, however, certain and expedient cost recovery and revenue assurances tend to favor investors over ratepayers and should be considered when authorizing rates of return.[49] This includes consideration of changes to the risk premium (reflecting the firm's sensitivity to systematic risks) as well as changes to the return premium (reflecting incentives associated with policy goals).

Risk and Rate Design

PRINCIPLE 28. **Rate design can shift revenue risk between utility investors and utility ratepayers and thus affect the utility's overall cost of capital.**

Increased reliance on fixed over variable charges tends to favor investors over ratepayers and should be considered when authorizing rates of return.

For public utilities, ratemaking generally involves three primary steps: the determination of revenue requirements (including the rate of return), the allocation of costs according to usage, and the design of rates charged (tariffs). Like rate of return, rate design involves a number of policy considerations and tradeoffs (table 13).

Revenue-neutral rate design means that rates should collect only authorized revenue requirements (no more or less). Efficient rates send accurate price signals and avoid undue price discrimination and wealth transfers, although regulators also consider equity and fairness in rate design. Utility bills generally are a composite of fixed charges that do not vary with usage (including a customer or meter charge) and variable charges (a rate applied to the level of usage). Traditionally, for most utilities, short-run fixed and variable costs are not matched to fixed and variable charges because some fixed costs are recovered through variable charges to preserve long-run price signals. As a consequence, revenues will vary with sales in proportion to volumetric rates (and sales will vary with a variety of risk factors).

Changing patterns of supply and demand across the utility industries are motivating interest in "straight fixed-variable" pricing that aligns costs and charges. In theory, doing so would make utilities indifferent to sales from an earnings perspective (a form of decoupling), but it also tends to reinforce a static view of infrastructure capacity. In the short term, many if not most utility costs are

TABLE 13. Tradeoffs between variable and fixed charges in rate design

RECOVERING MORE OF THE UTILITY'S REVENUE REQUIREMENT FROM FIXED CHARGES	RECOVERING MORE OF THE UTILITY'S REVENUE REQUIREMENT FROM VARIABLE CHARGES
▪ Enhances revenue stability for utilities (less sales risk)	▪ Reduces revenue stability for utilities (more sales risk)
▪ Weakens price signals and customer control (less resource efficiency)	▪ Strengthens price signals and customer control (more resource efficiency)
▪ Less affordable for low-income households (more regressive)	▪ More affordable for low-income households (less regressive)

effectively fixed (and possibly sunk). In the long run, of course, *all costs are variable*, and usage-neutral price signals will undermine ratepayer control and economic efficiency under dynamic conditions. Straight fixed-variable pricing may be especially problematic for the water industry, where variable pricing can be used to promote both efficiency and equity despite particularly high fixed costs.

Utilities are inclined to impose higher fixed charges in order to enhance revenue stability, which in turn stabilizes earnings. Revenue stability, of course, is but one of several criteria relevant to rate design (Bonbright et al. 1988). A greater reliance on fixed charges makes revenues less susceptible to short-run changes in usage (such as weather-related effects) but also long-run changes in preferences (such as technological developments and cultural shifts affecting overall usage patterns). Careful rate design, informed by cost-of-service studies and an understanding of price elasticities, offers some revenue stability and predictability even under variable pricing schemes. Maintaining lower fixed charges also helps address concerns about equity and affordability for ratepayers because household expenditures for utilities are generally regressive. Only large or combined households may see a slight advantage to greater reliance on fixed charges. New methods of rate design, including dynamic pricing to induce demand response, are promising in terms of improving both allocative efficiency (by aligning costs and prices) and capacity utilization (by shifting load). These methods also shift cost variability and related price risks to ratepayers, with distributional consequences (interclass and intraclass) depending on allocation rules. Customers with price-elastic usage and more capacity to respond to prices will be advantaged; customers with price-inelastic usage and less capacity to respond to prices will be disadvantaged. Long-term improvement in end-use efficiency will tend to narrow opportunities for price arbitrage. Net metering and rate design for customers who generate power also raise issues with regard to allocating costs and risks associated with electricity grids.

As more options emerge in relation to various policy goals, it is appropriate to recognize that choices about rate design (including rate structures and various mechanisms that flow through to rates) can shift risks in relevant ways. In particular, rate-design methods that neutralize revenue variability will enhance the utility's earnings potential. Thus, increased reliance on fixed over variable charges tends to favor investors over ratepayers and should be considered when authorizing rates of return. As in the use of cost-recovery and revenue-assurance mechanisms, this includes consideration of regulatory determination of both the risk and return premia (Principle 27).

Risk and the Cost of Service

PRINCIPLE 29. Regulatory policies that shift risk from utility investors to utility ratepayers may increase the overall cost of service.

Lower risk to investors may come at a high price to ratepayers, namely, an offsetting loss of economic efficiency due to weak performance incentives.

Collectively, regulatory policies define the regulatory environment for all stake-holders in the process. The prospect of creating a "responsive," "supportive," or "constructive" regulatory environment (see Moody's Investors Service 2014a) may have intuitive appeal, but a closer reading and an eye toward risk urges caution. Although perhaps well intentioned, many policies advanced toward this end shift the potential impacts of systematic and firm-specific risks from investors to ratepayers. In this respect, they tend to favor earnings over the overall cost of service.

Much of the focus on the regulatory environment comes from the utility credit-rating agencies, including Moody's and Standard and Poor's. These agencies should be largely indifferent to regulatory policy and its effect on individual utilities, as their role is simply to reveal the characteristics that distinguish credit quality based on risk, not influence regulatory policy. Unsurprisingly, however, credit analysts consistently identify certain methods for cost recovery, revenue assurance, and rate design (see Principles 27 and 28) with a positive (that is, credit-friendly) regulatory environment, particularly with respect to the relationship between regulatory lag and realized returns.[50] Regulatory certainty along these lines (that is, lower regulatory risk) is associated with positive credit ratings and lower costs of capital. Bondholders will favor regulatory methods that reduce cash-flow variability and thereby enhance the market value of bonds. Consistent with a bondholder

TABLE 14. Offsetting effects associated with lower risk and lower efficiency (hypothetical data)

CREDIT RATING = BBB

CAPITAL COSTS	CAPITAL STRUCTURE	AFTER-TAX COST OF CAPITAL	CAPITAL BASE	ANNUAL COST
DEBT	50%	3.0%	300,000	$9,000
EQUITY	50%	10.0%	300,000	30,000
				39,000
OPERATING COSTS				40,000
TOTAL REVENUE REQUIREMENTS				$79,000

CREDIT RATING = A

CAPITAL COSTS	CAPITAL STRUCTURE	AFTER-TAX COST OF CAPITAL	CAPITAL BASE	ANNUAL COST
DEBT	50%	2.8%	300,000	$8,400
EQUITY	50%	9.7%	300,000	29,100
				37,500
OPERATING COSTS WITH 10% COST OVER-RUN				44,000
TOTAL REVENUE REQUIREMENTS				$81,500

(as compared to shareholder) bias, some credit analysts have even acknowledged that "small declines in allowed ROE are less of a concern than timely recovery of costs" (Wobbrock 2014).

Still, regulators should be circumspect about utility credit ratings, as well as agency ratings, and the policies that enhance them. Lower risk to investors may come at a high price to ratepayers, namely, an offsetting loss of economic efficiency due to weak performance incentives. In other words, failing to meet the criteria for high ratings should not be construed as a sign of regulatory failure, at least not from the perspective of ratepayers. As long as investment-grade quality is maintained, lower ratings are unlikely to jeopardize capital attraction.

As demonstrated under Principle 15, firms with strong bond ratings based on a capital structure that limits debt may actually incur higher overall costs of capital. As illustrated in table 14, holding the capital structure constant, an improved credit rating (from BBB to A) could result in capital-cost savings of $1,500. However, if utility managers are less inclined to control operating costs because they can be easily passed along to ratepayers, a 10 percent cost increase could result in additional revenue requirements of $4,000 (a net effect of $2,500). In other words, lowering regulatory risk and shifting risk away from investors may have the unintended consequence of inflating total costs.

Thus, firms with strong bond ratings based on favorable regulatory policies may incur lower costs of capital but a higher overall cost of service. The reason is simply that utilities may become complacent and allow capital or operating costs to escalate above efficient levels, resulting in a net increase in revenue requirements that must be recovered from ratepayers. In addition to efficiency losses, as already noted, the complacent utility is less likely to innovate in ways that will enhance ratepayer value over the long term.

Risk and Sales Erosion

PRINCIPLE 30. **Regulation provides for the periodic adjustment of rates to account for changing usage, including erosion of sales.**

In the regulatory context, short-term revenue risk is a function of deviations from anticipated revenue requirements and anticipated sales.

Over time, evolving technologies and consumer preferences may be reflected in the erosion of sales for some services, even as other services grow in market share. The shift from landline to mobile communications services provides a stark example (figure 16). Over the long term, lack of sales growth is not a relevant risk for a utility's diversified equity investors unless it is associated with systematic macroeconomic forces affecting all firms across all sectors. In the short term, however, downward trends in utility sales may pose a challenge to bondholders and managers in terms of cash flows (see Principle 29). Some industry analysts question whether regulation and ratemaking methods can be adequately responsive to this challenge, which is primarily cast as a function of regulatory lag.[51]

Random variability in sales, due primarily to weather effects in the case of utilities, are inconsequential for diversified investors or the cost of equity. Energy and water utilities today are fretful about regulatory lag in the context of nonrandom variability in sales, namely declining sales in the face of high fixed and rising costs (see Kind 2013). Mandated end-use efficiency standards are sometimes characterized as a source of revenue risk for utilities because they contribute to reduced sales and thus sales revenues. Changes in usage that are driven by technical standards can be considered sector-specific, because they affect all firms in the sector. However, the associated investment risk is generally not relevant because

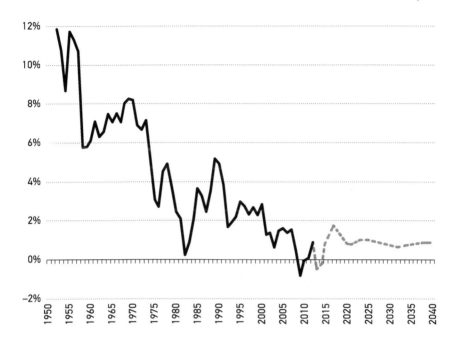

FIGURE 16. Annual percentage growth in electricity usage, 1950–2040

SOURCE: DATA FROM THE U.S. ENERGY INFORMATION ADMINISTRATION (WWW.EIA.GOV) WITH FORECAST FROM 2012 (INCLUDES SALES AND DIRECT USAGE).

it is still diversifiable. A portfolio holding manufacturers of high-efficiency fixtures (for example, light bulbs or toilets) along with utilities will experience offsetting effects. Even if the portfolio does not contain the manufacturers of high-efficiency fixtures, the random variability of the cash flows of unrelated entities across sectors will provide the diversification benefit and eliminates the impact of all but the systematic factors affecting all companies.

In addition, several forces mute the effects of end-use efficiency. Utilities are monopolistic, and utility services are essential and relatively (though not absolutely) price-inelastic. In the short term, end-use efficiency will lower sales, but it will also lower the utility's variable operating costs and revenue requirements, at least by some amount. Over the long term, end-use efficiency should actually lead to less volatile and more predictable usage that poses less risk in terms of cost recovery as well as capital planning.[52]

As conditions change, utilities can seek rate relief from regulators. In the regulatory context, short-term revenue risk is a function of deviations from anticipated revenue requirements (the numerator in ratemaking) and anticipated sales (the denominator in ratemaking). For some utilities, the disparity of authorized and realized returns may have as much to do with the disparity of forecast and realized sales than any other determinant, including regulatory policies concerning cost recovery and returns.

Regulators must recognize significant trends in sales when approving rates because rate revenues must provide utilities a reasonable opportunity to earn a fair return on their invested capital.[53] Regulators bear a special responsibility to recognize the effects of state-mandated and utility-sponsored programs that actively promote efficiency and conservation. Failure to adjust sales expectations may impair the utility's ability to cover short-run fixed costs. Furthermore, if that failure results in consistently subpar earnings for the utility, it might even be characterized as a constitutional "taking."

In setting rates, a moving average will not adequately estimate sales for the rate test year if the sales trend is nonrandom and trending downward. Prospective or forward-looking ratemaking (that is, future test years) is generally responsive. In addition, "demand-repression" adjustments may be used to calibrate the effects of price changes on price-elastic usage subsequent to a utility rate increase.[54] Rates can be adjusted for the effects of end-use efficiency on both revenues (sales forecasting) and operating costs (cost forecasting). As discussed under Principles 27 and 28, certain revenue-assurance mechanisms (such as revenue decoupling) and rate-design options (such as straight fixed-variable pricing) can further shift revenue risk from investors to ratepayers. All of these are discretionary policies that should be carefully considered within the larger context of a rate review that accounts for risk allocation.

Nonetheless, the utility should continue to bear a burden of proof in demonstrating "natural" and "programmatic" sales trends that are nonrandom and durable, a burden that can be challenging. The regulator should also ensure that avoided costs resulting from efficiency are promptly reflected in revenue requirements and rates to the benefit of ratepayers. Finally, the focus on sales revenue erosion and earnings, and ratemaking methods to address them, should not obscure attention to transformative trends in technologies and preferences that may permanently alter utility business models and investment profiles (discussed under Principle 31).

Risk and Investment Erosion

PRINCIPLE 31. Regulation does not protect utilities from endemic economic forces and market risks, including erosion of investment opportunity.

The regulatory compact and prevailing standards of review do not ensure that utilities will survive and thrive in perpetuity.

Although much is made of the short-term loss of utility sales volume, the more intractable problem for public utilities, individually and as a sector, might be the long-term loss of investment opportunity (figure 17). The obligation of regulation is to provide utilities with the opportunity to earn a fair return on the prudent and useful rate base, but it is not to preserve the rate base itself or the demand that justifies the associated capacity and investment. Put another way, it is only the percentage rate of return that is obligated, not the value to which it is applied or the total dollar returns that might be generated. With time, the proportionate market capitalization of the utility may actually decline.

Efficiency gains that result in loss of sales can help utilities avoid operating costs in the short run and capital costs in the long run. Indeed, avoiding costs is the central purpose of efficiency-oriented pricing and programs. In the context of declining sales, all utility investment decisions must be closely examined. The evaluation of prudence takes contemporary conditions into account. Given changing demand and environmental regulations, modular and flexible infrastructure design may be needed as part of an active risk-management strategy with regard to capital investment. Many utilities and their investors will need to adjust expectations and consider the potential movement away from a rate-base oriented growth paradigm.

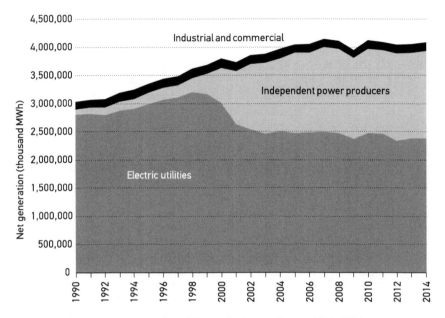

FIGURE 17. Electricity generation by utilities and other producers, 1990–2014

SOURCE: DATA FROM THE U.S. ENERGY INFORMATION ADMINISTRATION (WWW.EIA.GOV).

The right to compensation is protective of utility investors to a point, but not absolutely and not indefinitely. The regulatory compact and prevailing standards of review do not ensure that utilities will survive and thrive in perpetuity. Economic forces and disruptive technologies that provide new choices for consumers can transform or displace the sector and even destroy value (see Graffy and Kihm 2014). If the usage is elastic, permanent demand destruction may ensue. Assuming the survival of the utility, supply and demand must be reoptimized according to a new equilibrium.

Market trends (such as changes in commodity prices) and emerging technologies (such as power generation from renewable resources and energy storage) are important drivers of demand, even for regulated services. Public policy also plays a role in accelerating, decelerating, or mitigating the effects of transformative change. Environmental policies, such as those dealing with carbon pollution and climate change, exert a substantial impact on the utilities sector. Of course, economic regulatory policies have implications for whoever bears the associated risks. Certain revenue-assurance mechanisms, such as cost-inflation adjustments and decoupling, tend to insulate utilities from broader economic forces (systematic risks) affecting competitive firms. Since these risks are already reflected in the cost of equity, implementation of these mechanisms has the effect of shifting risk from investors to ratepayers (regarding inflation, see Kahn 1988, 115–16).

Technological transformation raises the possibility of market restructuring, but also stranded investment. For example, the need for high-voltage transmission might be lower in a less centralized energy future. Regulators may or may not be inclined to see the utility through a transformative period by, for example, ensuring revenue streams or even securitizing investment that would otherwise be stranded. Whether the regulatory compact implies an entitlement to recovery of stranded cost due to public policy forces is a matter of considerable debate (see Rose 1996).

As an institution, economic regulation ensures neither revenues ("coupled" or "decoupled") nor profitability nor even survival for public utilities, as they are known today. Nor should ratepayers bear risks associated with diversification by utilities into unregulated activities.[55] Regulators may be shield utilities from some risks up to a point, in order to serve social goals, but not to the extent of preserving a business model or operation that is not desirable or viable over the long term. When market or other social forces overwhelm the traditional utility model, traditional regulatory standards designed in accordance with that model no longer pertain. Particularly relevant here is the *Market St. Railway* (1945) case, where the Court recognized that it is not regulation's purpose to protect a business concern from impending demise brought on by emergent competition, evolving consumer preferences, and demand response to prices, where no regulated rate can ensure financial sustainability.[56] Extinction of companies and industries simply reflects market forces, which in turn reflects consumer sovereignty. Financial capital will flow accordingly, in some cases displacing regulated rate base with competitive enterprise marked by commensurate risks and returns.

Risk and Restructuring

PRINCIPLE 32. Restructuring and deregulation introduce considerable risk and uncertainty to public utilities and utility ratepayers.

Competitive markets obviate the path to profitability provided to public utilities by economic regulation.

Economic regulation is essential when essential services are provided by monopolies. The value of independent regulation as "enshrined in law" is recognized in the context of economic development, historically for the United States and for today's emerging economies.[57] But when economic and technological conditions allow for workable competition, regulation can be relaxed or eliminated. In other words, if and when competition is sufficiently workable to protect consumers and prevent abuse of market power, deregulation is feasible and appropriate (assuming tolerable market imperfections). Legislation and regulation inevitably play a role in the transition from regulated monopoly to competitive firm, ideally without favoring any particular technology, interest, or business model.

As emphasized above, the regulatory compact defines the rights and obligations of regulated utility service providers and various conditions of service. The compact in many respects is an "all or nothing" proposition. Despite variations in regulatory methods, a service or service provider is generally regulated or not. In other words, discipline for performance, including risk, is provided either by regulation or markets. Under some circumstances, and with sufficient safeguards in place, either competition or public ownership might serve the public interest better than economic regulation, even though regulation may have prevailed in the past (Principles 19 and 20).

TABLE 15. Restructuring and identification of credit supportive jurisdictions

STRUCTURAL STATUS	LEAST CREDIT SUPPORTIVE	LESS CREDIT SUPPORTIVE	CREDIT SUPPORTIVE	MORE CREDIT SUPPORTIVE	MOST CREDIT SUPPORTIVE
Regulated states	NM*	AZ* HI MS MO UT VT WA WV* WY	AR* CO FL ID IN KS KY LA MN NC ND NV* OK* SD VA*	AL GA IA SC WI	
Partial		MT	OR	CA MI	
Retail access	DE DC	CT IL ME MD NY RI TX	MA NH NJ OH PA		

*Retail access repealed. Excludes AK, NE, and TN.
Sources: Data on structural status from Kenneth Rose, Senior Fellow, Institute of Public Utilities; data on credit ratings from Standard & Poor's (2012b).

Thus, the choice to deregulate has consequences, including salient implications for risk (see Hierzenberger 2010). In competitive markets, risk flows directly from the marketplace. A firm that moves from a regulated to a competitive environment will be subject to a wide range of economic and other influences without the buffering effects of regulation, including traditional as well as alternative methods. Echoing various principles discussed here, to the extent that deregulation changes any risk, utility bondholders, managers, and ratepayers will be affected. To the extent that deregulation changes sensitivity to systematic risk, even diversified equity investors will be affected.

TABLE 16. Restructuring and identification of equity supportive jurisdictions

STRUCTURAL STATUS	HIGHEST RISK	HIGHER RISK	HIGH RISK	HIGH-MODERATE RISK	MODERATE RISK	LOW-MODERATE RISK	LOW RISK	LOWER RISK	LOWEST RISK
Regulated states					HI				
					ID				
					KS				
					MN	CO			
					MO	FL			
					NE	GA			
				AZ*	NV*	KY			
				AR*	OK*	LA		AL	
				VT	SD	ND	IN	MS	
				WA	UT	SC	IA	NC	
			NM*	WV*	WY	TN	VA*	WI	
Partial			MT	OR		CA			
						MI			
Retail access	CT	IL	TX	NH	DE	OH			
		MD		NY	DC				
				PA	ME				
				RI	MA				
					NJ				

*Retail access repealed. Excludes AK.
Sources: Data on structural status from Kenneth Rose, Senior Fellow, Institute of Public Utilities; data on equity ratings from Regulatory Research Associates (January 19, 2012) as reported in testimony by K. Norwood before the Washington Utilities and Transportation Commission (UE-120436 and UG-120437).

As a general proposition, competition shifts some risks away from ratepayers and toward investors with commensurate opportunities for reward. Deregulation may cause some firms to suffer losses, but other firms could thrive. Electricity market restructuring is largely credited with increased power-plant efficiency (Markiewicz, Rose, and Wolfram 2003), although these gains may be offset to some degree by the loss of scope economies and added transaction costs. Some believe that a regulated environment is more conducive to planning and coordination; others believe that a deregulated environment is more conducive to technological innovation, product differentiation, and alternative service-delivery models to enhance consumer value.

Although regulation can determine the cost of service, market forces ultimately determine the value of service.

Competitive markets obviate the path to profitability provided to public utilities by economic regulation. In other words, deregulated firms must find their own way. The greater uncertainty associated with competition likely explains why regulatory environment ratings from both the credit and equity perspectives appear to favor "supportive" regulated jurisdictions over jurisdictions with retail access (tables 15 and 16). Economic regulation provides a degree of stability and certainty for making large capital investments in the context of environmental mandates, and thus may be consistent with broader public-interest objectives.[58]

Deregulation transforms ratepayers into consumers, introducing opportunities and risks to them as well. Market restructuring in the energy sector is based largely on the premise that consumer choice will drive efficiency among competitive service providers. If a market is truly competitive, alternative providers should be able to meet consumer needs based on price, reliability, or other criteria. If reasonable alternatives for desired service levels are not available, then the market is not truly competitive. Market-determined prices, however, can be more volatile than regulated prices. Consumers can purchase protection against service disruption or price volatility, which adds to their costs and introduces the chance of regret. Absent established and enforced standards, markets can also introduce uncertainty about service quality and reliability due to supply-chain and capacity conditions but also due to competitor business practices. In addition, deregulation will have distributional effects (winners and losers) if interclass or intraclass transfers or subsidies were previously embedded in regulated prices. Depending on social goals, remedial policies may be needed to address these effects.

Risk Evaluation

PRINCIPLE 33. **Risk should be considered as only one of many factors within a broader evaluation framework.**

Reducing risk exposure involves tradeoffs with other decision criteria, including costs and economic efficiency.

Risk evaluation is challenging and complex. Indeed, a more rigorous approach to risk evaluation is much needed, especially in the regulatory context where risk allocation has important distributional and performance implications. Analyzing risk outside of a comprehensive framework could lead to unexpected or unwanted outcomes.

Colloquial conceptions of risk feed the impulse toward risk avoidance. Regulators have been encouraged to "seek to identify, understand and minimize the risks associated with electric utility resource investment" (Binz et al. 2012). The implications of this recommendation for resource planning and policy are considerable, but two more fundamental concerns arise. First, as noted throughout this analysis, risk is simply the probability of being wrong either on the upside or the downside. The potential for upside gains can rationalize taking on some amount of risk. Second, optimal decision making calls for assessing risk within a larger evaluation framework that includes criteria other than risk. In other words, risk is only one input, characteristic, or evaluation criterion. Decision makers would be remiss to consider only risk when assessing their options.

Reducing risk exposure, then, involves tradeoffs with other decision criteria, including costs and economic efficiency. As already noted, shifting risk can alter incentives, and avoiding risk can result in unintended consequences. The goals of reliability and security for utility services provide a simple yet powerful example

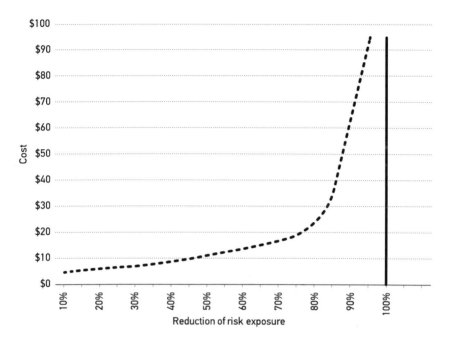

FIGURE 18. Effect of reducing risk exposure on cost (hypothetical data)

(figure 18). As a specific example, all electricity distribution infrastructure could be located underground to prevent outages from downed trees, but the expense to ratepayers would be extremely high. Absolute reliability is elusive, and pursuing it would be very costly in nominal terms but also in terms of economic efficiency (that is, society incurs a welfare loss). Minimizing the chance of any outage will invariably fail to minimize costs or maximize efficiency. Given implications for access, affordability, and acceptability, the pursuit of absolute reliability can lead to unnecessary expense and considerable regret.

Risk reduction is popularly associated with favorable credit ratings and a lower cost of debt capital, and thus a lower cost of service. Regulators are encouraged to provide assurances to bond markets, and rewarded with positive ratings for their agencies in the process. Although it might seem like a "no brainer," a more holistic perspective should give pause. Specifically, the marginal savings associated with debt costs might be more than offset by efficiency losses and foregone innovation associated with weak performance incentives. The semblance of guaranteed returns can lull utilities into complacency, at a price to society measurable in terms of opportunity costs.

Damodaran (2008, 9) emphasizes that "businesses that are in a constant defensive crouch when it comes to risk are in no position to survey the landscape and find risks they are suited to take." Recent decades suggest that utilities have become notably risk-averse, perhaps reinforced by regulatory policy. Taking on unnecessary risk is never advisable, but neither is risk avoidance. Decision tools are available to private and public decision makers for strategic risk management (including insurance and hedging). Within risk tolerances and other constraints, risk can be optimized relative to other criteria.

As seen through a wider lens, society advances beyond the status quo by taking calculated risks, not by shunning risk. Put differently, risk avoidance comes at a price. Risk that offers the chance of returns motivates innovation in regulated and competitive businesses alike. As stewards of the public interest, policymakers, like investors, should seek not to minimize risk but to maximize value, considering risk tolerance and relevant evaluation criteria other than risk.

Conclusions

Risk can be poorly understood in general and in the realm of public utility regulation. Common usage of the term is invariably pejorative and imprecise. When risk is equated with the possibility of harm, and only harm, risk avoidance appears to be the prudent response. Yet to avoid all risk is to accept stagnation over progress.

Risk literacy seems more important than ever, as several important truths about risk are rarely articulated. A rhetorical understanding of risk is insufficient at best and harmful at worst. A firm grasp of risk principles is essential to critical thinking about risk as well as to regulatory practice with regard to risk allocation. These principles remain constant, even in a context of substantial technological and policy change. In fact, dynamic conditions call for a more robust treatment of risk.

By applying the principles elucidated here, the concept of a risk proper and its various dimensions should be comprehensible. Investors and entrepreneurs know that risks present both downsides and upsides, and that value is created not by risk avoidance but rather by strategic risk-taking and risk management. Finance theory emphasizes that portfolio diversification is key to managing risk exposure.

Risk is a matter of perspective, and some risks will be relevant to the beholder while others will not. Many risks that matter a great deal to corporate managers and somewhat to bondholders are irrelevant to diversified equity investors. Information from only one perspective may have limited value in risk assessment. Decisions that minimize risk may actually increase the chance of regret, raise total costs, and thwart efficiency and innovation.

The regulatory compact, tested over more than a century, provides substantial guidance about risk and its allocation. Regulation, substituting for competition, is meant to maintain the connection between risk and reward. Regulators generally authorize fair rates of return that exceed the cost of capital to motivate socially beneficial investment. Any return premia (fair returns as well as bonus returns) can and should be tied to explicit performance goals. Adaptive regulatory policies that ostensibly reduce risk to utilities typically have the effect of shifting risk to ratepayers. Although regulation must be responsive to short-term financial challenges, the compact does not ensure that utilities will survive in perpetuity. For both utilities and consumers, restructuring and deregulation trade regulatory risk for market risk.

The emotional response to uncertainty appears to drive the impulse to reduce risk exposure. Risk avoidance engenders complacency and perpetuates the status quo. Ironically, uncertain times call for more, not less risk taking. Today, as in the past, properly allocated risk can motivate public utilities toward economic progress. From society's perspective, avoiding risk may present the greatest risk of all.

Beta Coefficients for Selected Regulated, Formerly Regulated, and Related Market Sectors (2015)

INDUSTRY	NUMBER OF FIRMS	LEVERED BETA	UNLEVERED BETA	D/E RATIO	STANDARD DEVIATION OF EQUITY
Telecom equipment	126	1.24	1.12	11.7%	62.74%
Reinsurance	4	1.35	1.02	37.6%	21.20%
Environmental and waste services	103	1.28	0.93	40.6%	65.61%
Transportation (railroads)	10	1.05	0.90	20.2%	30.73%
Trucking	30	1.32	0.89	66.7%	48.49%
Oil and gas (production and exploration)	392	1.27	0.87	48.2%	71.93%
Broadcasting	28	1.30	0.82	71.1%	62.12%
Coal and related energy	42	1.64	0.78	110.4%	74.62%
Insurance (general)	24	1.03	0.76	43.0%	35.35%
Utility (water)	19	1.09	0.76	50.2%	38.21%
Oil/gas (integrated)	8	0.81	0.74	11.1%	33.56%
Transportation	21	0.86	0.73	21.0%	42.36%
Telecom services	77	1.07	0.68	65.4%	55.60%
Insurance (life)	25	1.04	0.68	71.2%	34.48%
Cable television	18	0.91	0.67	44.8%	52.83%
Total market	7,887	1.06	0.67	66.1%	53.60%
Oil and gas distribution	85	0.96	0.66	47.8%	43.15%
Insurance (property and casualty)	52	0.83	0.66	32.9%	36.73%
Investments and asset management	148	1.10	0.65	74.0%	41.63%
Green and renewable energy	26	1.32	0.63	110.0%	53.18%
Air transport	22	0.98	0.59	81.5%	53.32%
Power	82	0.83	0.52	75.7%	29.70%
Telecom (wireless)	21	1.15	0.47	154.4%	53.05%
Utility (general)	21	0.59	0.42	61.3%	23.02%
Banks (regional)	676	0.53	0.33	77.69%	37.41%
Brokerage and investment banking	46	1.16	0.32	303.5%	44.77%
Bank (money center)	13	0.81	0.31	218.7%	39.98%
Financial services (nonbank and insurance)	288	0.67	0.06	1,206.7%	38.80%

Source: Damodaran Online (http://www.stern.nyu.edu/~adamodar/pc/datasets/betas.xls).

Key Court Decisions Regarding Risk

Public utility regulation rests firmly on a series of judicial decisions, including several handed down by the U.S. Supreme Court. In this appendix, we highlight some of the fundamental cases related directly to the establishment and application of risk principles.

Smyth v. Ames, 169 U.S. 466 (1898)

"What the company is entitled to ask is a fair return upon the value of that which it employs for the public convenience, and on the other hand, what the public is entitled to demand is that no more be exacted from it for the use of a public highway than the services rendered by it are reasonably worth." Reasonable rates are based on the "fair value" of the rate base and returns are not guaranteed: "The public cannot properly be subjected to unreasonable rates in order simply that stockholders may earn dividends. . . . If a corporation cannot . . . earn dividends for stockholders, it is a misfortune for it and them which the Constitution does not require to be remedied by imposing unjust burdens upon the public." Overturned by *FPC v. Natural Gas Pipeline* (1942).

Cedar Rapids Gas Light Co. v. City of Cedar Rapids, 223 U.S. 655 (1912)

"On the one side, if the franchise is taken to mean that the most profitable return that could be got, free from competition, is protected by the Fourteenth Amendment, then the power to regulate is null. On the other hand, if the power to regulate withdraws the protection of the Amendment altogether, then the property is nought. This is not a matter of economic theory, but of fair interpretation of a bargain. Neither extreme can have been meant. A midway between them must be hit."

Bluefield Waterworks v. PSC of WV, 262 U.S. 679 (1923)

"Rates which are not sufficient to yield a reasonable return on the value of the property used at the time it is being used to render the service of the utility to the public are unjust, unreasonable, and confiscatory, and their enforcement deprives the public utility company of its property, in violation of the Fourteenth Amendment.... A public utility is entitled to such rates as will permit it to earn a return on the value of the property it employs for the convenience of the public equal to that generally being made at the same time and in the same region of the country on investments in other business undertakings which are attended by corresponding risks and uncertainties, but it has no constitutional right to profits such as are realized or anticipated in highly profitable enterprises or speculative ventures.... The return should be reasonably sufficient to assure confidence in the financial soundness of the utility, and should be adequate, under efficient and economical management, to maintain its credit and enable it to raise the money necessary for the proper discharge of its public duties.... A rate of return may be reasonable at one time, and become too high or too low by changes affecting opportunities for investment, the money market, and business conditions generally."

Southwestern Bell v. PSC of Missouri, 262 U.S. 276 (1923)

"What will amount to a fair return cannot be ascertained by valuing the property as of past times without giving consideration to greatly increased costs of labor, supplies, etc., prevailing at the time of the investigation.... A state Commission, in fixing the rates of a public utility corporation, cannot substitute its judgment for the honest discretion of the company's board of directors respecting the necessity and reasonableness of expenditures made in the operations of the company.... The term 'prudent investment' is not used in a critical sense. There should not be excluded from the finding of the base investments which, under ordinary circumstances, would be deemed reasonable. The term is applied for the purpose of excluding what might be found to be dishonest or obviously wasteful or imprudent expenditures. Every investment may be assumed to have been made in the exercise of reasonable judgment unless the contrary is shown."

Denver Union Stock Yard Co. v. U.S., 304 U.S. 470 (1938).

"As of right safeguarded by the due process clause of the Fifth Amendment, appellant is entitled to rates, not per se excessive and extortionate, sufficient to yield a reasonable rate of return upon the value of property used at the time it is being used, to render the services. But it is not entitled to have included any property not used and useful for that purpose."

Federal Power Commission v. Natural Gas Pipeline Co., 315 U.S. 575, 585 (1942)

"By longstanding usage in the field of rate regulation, the 'lowest reasonable rate' is one which is not confiscatory in the constitutional sense. Assuming that there is a zone of reasonableness within which the Commission is free to fix a rate varying in amount and higher than a confiscatory rate, the Commission is also free under § 5(a) to decrease any rate which is not the 'lowest reasonable rate.' ... But regulation does not insure that the business shall produce net revenues, nor does the Constitution require that the losses of the business in one year shall be restored from future earnings by the device of capitalizing the losses and adding them to the rate base on which a fair return and depreciation allowance are to be earned."

Federal Power Commission v. Hope Natural Gas Co., 320 U.S. 591 (1944)

"Under the statutory standard of 'just and reasonable,' it is the result reached, not the method employed, which is controlling. . . . From the investor or company point of view, it is important that there be enough revenue not only for operating expenses, but also for the capital costs of the business. These include service on the debt and dividends on the stock. By that standard, the return to the equity owner should be commensurate with returns on investments in other enterprises having corresponding risks. That return, moreover, should be sufficient to assure confidence in the financial integrity of the enterprise, so as to maintain its credit and to attract capital. . . . The ratemaking process under the Act, i.e., the fixing of 'just and reasonable' rates, involves a balancing of the investor and the consumer interests. . . . Rates which enable the company to operate successfully, to maintain its financial integrity, to attract capital, and to compensate its investors for the risks assumed certainly cannot be condemned as invalid, even though they might produce only a meager return on the so-called 'fair value' rate base. . . . The fact is that, in natural gas regulation, price must be used to reconcile the private property right society has permitted to vest in an important natural resource with the claims of society upon it—price must draw a balance between wealth and welfare."

Market St. Railway Co. v. Railroad Commission of California, 324 U.S. 548 (1945)

"The problem of reconciling the patrons' needs and the investors' rights in an enterprise that has passed its zenith of opportunity and usefulness, whose investment already is impaired by economic forces, and whose earning possibilities are already invaded by competition from other forms of transportation, is quite a different problem. . . . the due process clause never has been held by this Court to require a commission to fix rates on the present reproduction value of something no one would presently want to reproduce, or on the historical valuation of a property whose history and current financial statements showed the value no longer to exist, or on an investment after it has vanished, even if once prudently made, or to maintain the credit of a concern whose securities already are impaired. The due process clause has been applied to prevent governmental destruction of existing economic values. It has not and cannot be applied to insure values, or to restore values that have been lost by the operation of economic forces. . . . Even monopolies must sell their services in a market where there is competition for the consumer's dollar and the price of a commodity affects its demand and use."

Federal Power Commission v. Sierra Pacific Power Co., 350 U.S. 348 (1956)

"While it may be that the Commission may not normally impose upon a public utility a rate which would produce less than a fair return, it does not follow that the public utility may not itself agree by contract to a rate affording less than a fair return or that, if it does so, it is entitled to be relieved of its improvident bargain. . . . In such circumstances, the sole concern of the Commission would seem to be whether the rate is so low as to adversely affect the public interest—as where it might impair the financial ability of the public utility to continue its service, cast upon other consumers an excessive burden, or be unduly discriminatory. . . . It is clear that a contract may not be said to be either "unjust" or "unreasonable" simply because it is unprofitable to the public utility."

El Paso Natural Gas Co. v. FPC, 281 F.2d 567 (1960)

"It is the obligation of all regulated public utilities to operate with all reasonable economies. This applies to tax savings as well as economies of management. The net result of this, of course, it that such savings as are effected are passed on to the consuming public. This we consider to be the natural and necessary consequence of rate regulation. . . . It is the duty of the Commission to fix a rate that represents the cost of service and a reasonable return on the investment, including compensation for consumption of the gas and an increment for incentive."

Trans World Airlines v. Civil Aeronautics Board, 385 F.2d 648 (1967)

"[The] issue is not whether the company acted lawfully but whether it acted prudently—a higher standard. . . . We are not unaware that the difficulties may be greater in practice than in philosophy in avoiding an improper usurpation of managerial discretion while conducting a proper review of abuse of that discretion, and that these difficulties are not lessened when Government officials have the 20–20 vision of hindsight."

Permian Basin Area Rate Cases, 390 U.S. 747, 790 (1968)

"Courts are without authority to set aside any rate selected by the Commission which is within a 'zone of reasonableness' . . . . No other rule would be consonant with the broad responsibilities given to the Commission by Congress; it must be free, within the limitations imposed by pertinent constitutional and statutory commands, to devise methods of regulation capable of equitably reconciling diverse and conflicting interests. It is on these premises that we proceed to assess the Commission's orders. . . . Neither law nor economics has yet devised generally accepted standards for the evaluation of rate-making orders."

Narragansett Elec. Co. v. Burke, 381 A. 2d 1358, Rhode Island Supreme Court (1977)

"In view of the statutory requirement, we do not believe that the delay by the PUC in ruling on the motion was that type of 'grossly excessive' delay which we have indicated may entitle the utility company to retroactive relief. It is a fundamental rule that utility rates are exclusively prospective in nature. Additionally, there is a presumption that current commission orders are valid. Therefore, absent extraordinary circumstances, the utility company must bear the risk of loss inherent in the well-known lag accompanying the making of rate changes."

Duquesne Light v. Barasch, 488 U.S. 299 (1989)

"The risks a utility faces are in large part defined by the rate methodology, because utilities are virtually always public monopolies dealing in an essential service, and so relatively immune to the usual market risks. Consequently, a State's decision to arbitrarily switch back and forth between methodologies in a way which required investors to bear the risk of bad investments at some times while denying them the benefit of good investments at others would raise serious constitutional questions. But the instant case does not present this question."

For more information, see Beecher 2014, available at ipu.msu.edu.

Glossary

Ambiguity. An attribute of the quality of information.

Beta. A measure of the sensitivity of an investment to systematic risk.

Capital structure. The combination of debt, equity, and other financial instruments (such as leases) deployed by private companies for financing capital investments and securitizing earnings.

Cost of capital. The cost of debt and equity, weighted by their respective levels and rates, deployed by companies for financial capital investments.

Cost of equity. The cost of financial capital supplied by shareholder investors.

Coverage. The capacity of a firm to meet its financial obligations from periodic earnings.

Credit risk. A firm-specific risk concerning the probability of a default on debt; also known as credit-default risk.

Diversification. A risk-management strategy involving concurrent holding of investments whose returns move with less than perfect positive correlation.

Effective risk. Overall risk adjusted for the covariance among multiple variables or risk factors.

Equity risk. Any risk shareholders cannot diversify away in a portfolio.

Financial risk. Equity risk associated with varying capital structures.

Firm-specific risk. Risk relevant only to one firm; also known as idiosyncratic risk.

Hedging. A strategy for changing the exposure to risk by offsetting risky transactions, transferring risks to other investors.

Interest-rate risk. A systematic risk concerning the probability of a change in interest rates.

Lag (regulatory). The delay between a change in costs or revenues (+/–) and a change in authorized prices.

Leverage. The proportion of debt or debt-like instruments (such as leases) in the company's capital structure.

Normal distribution. Where possible outcomes are distributed symmetrically around the average, and the dispersion around the average is measured completely by the standard deviation from the mean.

Option. A real or financial security granting the right to trade (buy with a call, sell with a put) in an underlying real or financial asset; also known as a call or put. Options truncate the distribution of returns to the underlying asset.

Regret. An emotion experienced after the fact when comparing outcomes; fear of regret can motivate behavior.

Risk. A probabilistic estimation of potential outcomes based on knowable information; in finance, the chance of deviation from expected returns.

Risk aversion. Behavior suggesting a preference against variable outcomes without commensurate compensation.

Risk avoidance. Behavior suggesting a preference against risk-taking regardless of compensation.

Risk factor. A discrete manifestation of risk or a "risk."

Risk management. A strategy for reducing risk exposure, mitigating risk effects, or shifting risk to others.

Risk premium. The additional return over a risk-free rate required to compensate investors for equity risk; also known as equity-risk premium or equity premium.

Sector-specific risk. Nonrandom risk affecting groups of like investment securities within the overall market in like ways.

Skew. A measure of asymmetry in the distribution of risky outcomes.

Systematic risk. Risk affecting an entire market or financial system; also known as market risk.

Uncertainty. A lack of knowledge that thwarts probabilistic estimation of future outcomes.

Volatility. The dispersion of returns over a period of time, typically measured in terms of standard deviation.

Notes

1. Time plays an important role in the perception of risk, particularly when measured by variability or volatility. Short-term patterns may appear very different when viewed through the wider lens that time affords.
2. "Uncertainty must be taken in a sense radically distinct from the familiar notion of risk, from which it has never been properly separated. . . . The essential fact is that 'risk' means in some cases a quantity susceptible of measurement, while at other times it is something distinctly not of this character; and there are far-reaching and crucial differences in the bearings of the phenomena depending on which of the two is really present and operating. . . . It will appear that a measurable uncertainty, or 'risk' proper, as we shall use the term, is so far different from an unmeasurable one that it is not in effect an uncertainty at all" (Knight 1921).
3. Ignored for the moment is the asymmetrical and thus incomplete presentation of risk in this particular model (that is, the exclusive consideration of downside risk in table 1).
4. Diet and exercise, for example, may reduce an individual's endogenous risk of experiencing certain health problems.
5. A well-worn example is that the 50–50 probability of outcome for a coin toss is unaltered by the number of tosses.
6. For more information, see Dixit and Pindyck (1994) and Amram and Kulatilaka (1999).
7. For a practical guide, see PricewaterhouseCoopers (2008).
8. Renewable energy resources do not carry the fuel-price risk of energy commodities. In addition, fuel-cost adjustment mechanisms approved in many regulatory jurisdictions shift at least some fuel-price risk to ratepayers.
9. Bean and Hoppock assert, "A least-risk metric that also assures low relative costs by 'minimizing the maximum regret' of generation plans is a potentially attractive alternative approach" (2013, 5). We disagree because least-risk options are often more expensive than high-risk alternatives, and choosing low-risk options can lead to considerable *ex post* regret. For example, if a utility wanted to minimize the chance of a power outage, it could increase its reserve margin, install redundant capacities, or even

provide customers with backup generators. The lights will be more likely to stay on, reducing the risk of an outage, but at a very high (and arguably imprudent) cost.

10. The all-stock portfolio outperformed the all-Treasuries portfolio in each of the 66 overlapping periods across the 1928–2012 time frame.

11. Higgins (1988, 230) makes the case that net present value is the superior financial metric for evaluating competing investment opportunities.

12. Evidence suggests that for most mergers, the shareholders of the acquired companies capture significantly more value than do those of the acquiring company (see Koller et al. 2010).

13. If a utility manager, for example, holds proprietary information and/or can persuade regulators that the company's risk profile is higher than it is in reality, then shareholders will realize returns in excess of those required by systematic risks.

14. Regulators often authorize returns on equity in excess of the cost of equity for policy purposes. But as we discuss later, this is a *return* premium unrelated to risk (see Principle 22).

15. Public Service Company of New Hampshire (1988); El Paso Electric (1992); Pacific Gas and Electric (2001); and Columbia Gas (1991).

16. The returns in the analysis consider capital gains and losses only. Most of the variation in stock returns flows from those gains and losses, not changes in dividend payments.

17. Technically, it is both the direction and scale of deviation from the average that determine price correlation and thus portfolio impact.

18. The example is more than conjecture. In the 2013–2014 period, following the introduction of the Android phone, Google saw its stock price rise by about 20 percent as Apple's declined by about the same amount.

19. Based on data from Yahoo Finance, the losses in market value for stocks by sector for this period were as follows: Financials (–71%), Industrials (–57%), Materials (–55%), Energy (–49%), Consumer Discretionary (–45%), Technology (–38%), Utilities (–35%), Health Care (–29%), and Consumer Staples (–28%). By comparison, Treasury bills produced no losses during this period.

20. On black swans and the effects of non-normal distributions on decision making, see Taleb (2010); see also Kahneman and Tversky (2000) and Gigerenzer (2014).

21. *Federal Power Commission v. Hope Natural Gas Co.*, 320 U.S. 591 (1944).

22. Of course, risk is complex and dynamic, so changes in systematic risk or the firm's sensitivity to systematic risk may simultaneously affect the cost of equity and thus stock prices.

23. "The risk that companies must identify and manage is their cash flow risk, meaning uncertainty about their future cash flows. Finance theory is, for the most part, silent about how much cash flow risk a company should take on. In practice, however, managers need to be aware that calculating expected cash flows can obscure material risks capable of jeopardizing their business when they are deciding how much cash flow risk to accept" (Koller, Goedhart, and Wessells 2010, 34–35).

24. Beta is a reasonable and accepted measure of risk, despite controversy over methodologies used in the conversion to required return (see Ross 1993).

25. The betas shown in figure 9 and appendix 1 are raw (unadjusted) betas. Adjusting betas for regression tendencies does not change the observation regarding the relative risk of the sectors.

26. *Bluefield Waterworks v. PSC of WV*, 262 U.S. 679 (1923) and *Federal Power Commission v. Hope Natural Gas Co.*, 320 U.S. 591 (1944).

27. To analogize, when circumstances improve for homeowners (such as an increase in income or home value), they do not offer to raise the interest rate on their mortgage debt.

28. Bondholders can protect themselves to some extent by deploying covenants and other financial instruments that limit the cost of unsystematic losses by placing them on investors who can diversify these risks. These instruments cannot help bondholders in terms of portfolio returns, but they can prevent risk-shifting *within the firm* from equity holders to bondholders. Again, however, what drives the cost of capital for both debt and equity is the effect changing risks *across firms*. Firm-specific risks of unrelated firms cancel out in equity portfolios because they move independently. The bondholders of a successful company experience no effect on the value of their bonds, but the bondholders of a struggling company will experience the effects of default risk in terms of lower bond prices. Covenants may prevent shareholders of the struggling company from shifting risk to bondholders, but the bond prices of the successful firm will be unaffected. Thus, in-firm covenants do nothing to address the risk associated with asymmetric treatment of good and bad news for debt holders.

29. According to Pratt and Grabowski (2010, 51), over the past eighty years the standard deviation in returns for the S&P 500 portfolio (where firm-specific risks cancel out) is about 21 percent, more than twice that experienced for a long-term corporate bond portfolio (where firm-specific risks do not cancel out).

30. With time, of course, disruptive technologies might provide competitive choices to consumers that can resolve the problem of market power and thus justify restructuring and deregulation.

31. This also raises principal-agency issues beyond the scope of this discussion.

32. On the public interest and the legislative purpose of utilities, see *Nebbia v. New York*, 291 U.S. 502 (1934).

33. Peltzman (1976, 230) notes the "buffering" effect of regulation with respect to economic forces, which reduces the effects of both systematic and firm-specific risk.

34. A certificate of need or "public convenience and necessity" is recognition of need but not a guarantee of cost recovery under the prudence standard.

35. Federal loan guarantees for nuclear-plant construction illustrate socialized risk borne by taxpayers.

36. In reality, the social optimum is elusive. Market forces, by definition, do not consider externalities (positive or negative). Moreover, while competitive pressure pushes prices toward marginal cost and returns toward the cost of capital, market power or other imperfections may allow some firms to earn supernormal profits for long periods.

37. The takings clause was established for the federal government under the Fifth Amendment and extended to the states under the Fourteenth Amendment's due process clause pursuant to *Chicago Burlington and Quincy R.R. v. City of Chicago*, 166 U.S. 226 (1897).

38. See *Federal Power Commission v. Natural Gas Pipeline Co.*, 315 U.S. 575, 585 (1942) and *Federal Power Commission v. Hope Natural Gas Co.*, 320 U.S. 591 (1944).

39. This implies logically that market-to-book ratios for regulated utilities would generally be above 1.0, although poor performing utilities could see market-to-book ratios below 1.0.

40. "A public utility is entitled to such rates as will permit it to earn a return on the value of the property which it employs for the convenience of the public equal to that generally being made at the same time and in the same general part of the country on investments in other business undertakings which are attended by corresponding risks and uncertainties, but it has no constitutional right to profits such as are realized or anticipated in highly profitable enterprises or speculative ventures." *Bluefield Waterworks v. PSC of WV*, 262 U.S. 679 (1923).

41. In other words, an investment that earns a return equal to the cost of capital produces a net present value of $0 for investors.

42. *Narragansett Elec. Co. v. Burke*, 381 A.2d 1358 (1977).

43. This definition is known as Hicksian income.

44. See *Trans World Airlines v. Civil Aeronautics Board*, 385 F.2d 648 (1967) and *Duquesne Light v. Barasch*, 488 U.S. 299 (1989).

45. For example, Virginia statutes (§ 56–585.1) provide for incentives based on "generating plant performance, customer service, and operating efficiency," as well as investment in alternative power generation technologies.

46. See "Network Regulation—the RIIO Model," available at OFGEM, www.ofgem.gov.uk.

47. In the 1970s and early 1980s, authorized returns below the cost of equity in the context of a substantial plant-investment cycle destroyed investor value and confidence.

48. Utility managers and their bondholders might embrace end-use efficiency of their own accord, due to its potential to reduce cash-flow variability associated with sales. In the municipal credit context, see Standard and Poor's (2012a, 37).

49. Evidence from decoupling proceedings between 2005 and 2012 suggests that regulators adjust authorized equity returns downward in fewer than 20 percent of cases (Morgan 2013).

50. "Although many companies are facing downward pressure on their return on equities (ROEs), we see evidence of regulatory authorities' willingness to provide special recovery mechanisms that reduce regulatory lag and enhance a utility's ability to earn its allowed level of ROE" (Moody's Investors Service 2013).

51. "Growth rates in unit sales of water, electricity, and natural gas for residential and commercial customers have fallen and in some regions have been negative. Electric consumption grew at less than 0.5% during 2000–2010. Natural gas consumption stagnated back in the 1970s and had no growth during 2000–2010. Public supply water consumption per capita declined from 1990–2005. This has taken away a source of funds for future investments and for overcoming regulatory lag that is built into the regulatory process" (Wharton et al. 2013, ES).

52. Reducing outdoor water usage, for example, would reduce the effects of sales revenue variability on water utilities.

53. Once rates are approved, subsequent regulatory lag will continue to provide incentives for cost control.

54. The interrelated nature of rising prices (to cover fixed costs) and falling demand (for price-elastic usage) is sometimes referred to as a "death spiral." The merits of this characterization are beyond the scope of this discussion.

55. The controversial issue of whether utilities should engage in competitive services is beyond the scope of this discussion.

56. "Transportation history of San Francisco follows a pattern not unfamiliar. This property has passed through cycles of competition, consolidation, and monopoly, and new forms of competition; it has seen days of prosperity, decline, and salvage. . . . The Commission found, however, that the service had constantly deteriorated, and was worse under the seven-cent fare than under the former five-cent rate. . . . Reviewing the financial results of fare increases, the Commission concluded that the Company would reap no lasting benefit from rates in excess of five cents, due to the tendency of a higher rate to

discourage patronage." *Market St. Railway Co. v. Railroad Commission of California*, 324 U.S. 548 (1945).

57. "The support of independent, stable, and transparent regulatory frameworks, or frameworks enshrined in law, reduce the risk of adverse policy changes on a transaction" (Standard & Poor's 2014).

58. "Regulated utilities, even those with heavy coal-fired generation fleets, will fare better than unregulated power generators. These utilities typically have established revenue mechanisms in place to recover any costs or investments required to comply with environmental mandates" (Moody's Investors Service 2014c).

References

Amram, M., and N. Kulatilaka. 1999. *Real Options: Managing Strategic Investment in an Uncertain World*. Watertown, MA: Harvard Business School Press.

Asness, C. 2014. "My Top 10 Peeves." *Financial Analysts Journal* 70, no. 1: 22–30.

Averch, H., and L. L. Johnson. 1962. "Behavior of the Firm under Regulatory Constraint." *American Economic Review* 52, no. 5: 1052–69.

Bean, P., and D. Hoppock. 2013. *Least-Risk Planning for Electric Utilities*. Working paper, Nicholas Institute, Duke University.

Beecher, J. A. 2013. "Economic Regulation of Utility Infrastructure." In *Infrastructure and Land Policies*, ed. G. K. Ingram and K. L. Brandt. Cambridge, MA: Lincoln Institute of Land Policies.

———. 2015. *Primer on Core Case Law in U.S. Public Utility Regulation*. East Lansing: Institute of Public Utilities, Michigan State University.

Binz, R., R. Sedano, D. Furey, and D. Mullen. 2012. *Practicing Risk-Aware Electricity Regulation: What Every State Regulator Needs to Know*. Boston, MA: CERES.

Bodie, Z. 1995. "On the Risks of Stocks in the Long Run." *Financial Analysts Journal* 51, no. 3: 18–22.

Bonbright, J. C., A. L. Danielsen, and D. R. Kamershen. 1988. *Principles of Public Utility Rates*. Arlington, VA: Public Utilities Reports.

Brealey, R. A., and S. C. Myers. 1984. *Principles of Corporate Finance*. New York: McGraw-Hill.

Brealey, R. A., S. C. Myers, and F. Allen. 2006. *Principles of Corporate Finance*. New York: McGraw-Hill Irwin.

Breyer, S. G. 1982. *Regulation and Its Reform*. Cambridge, MA: Harvard University Press.

Brigham, E. F., L. C. Gapenski, and D. A. Aberwald. 1987. "Capital Structure, Cost of Capital and Revenue Requirements." *Public Utilities Fortnightly* 119, no. 1: 15–24.

Brilliant, H., and E. Collins. 2014. *Why Moats Matter: The Morningstar Approach to Stock Investing*. Hoboken, NJ: John Wiley.

Canterbery, E. R., B. Johnson, and D. Reading. 1996. "Cost Savings from Nuclear Regulatory Reform: An Econometric Model." *Southern Economic Journal* 62, no. 3: 554–66.

Clarke, R. G. 1980. "The Effect of Fuel Adjustment Clauses on the Systematic Risk and Market Values of Electric Utilities." *Journal of Finance* 35, no. 2: 347–58.

Congressional Budget Office. 1986. *The Financial Condition of the U.S. Electric Utility Industry.* Washington, DC: U.S. Government Printing Office.

D'Ambrosio, C. 1990. "Portfolio Management Basics." In *Managing Investment Portfolios: A Dynamic Process*, ed. J. L. Maginn and D. Tuttle. 2nd ed. Boston, MA: Warren Gorham and Lamont.

Damodaran, A. 2001. *Corporate Finance: Theory and Practice.* 2nd ed. Hoboken, NJ: John Wiley and Sons.

———. 2008. *Strategic Risk Taking.* Upper Saddle River, NJ: Pearson Education.

———. 2011. *Applied Corporate Finance.* 3rd ed. Hoboken, NJ: John Wiley and Sons.

De Fraja, G., and C. Stones. 2004. "Risk and Capital Structure in the Regulated Firm." *Journal of Regulatory Economics* 26, no. 1: 69–84.

Diermeier, J. 1990. "Capital Market Expectations: The Macro Factors." In *Managing Investment Portfolios: A Dynamic Process*, ed. J. L. Maginn and D. Tuttle. 2nd ed. Boston, MA: Warren Gorham and Lamont.

Dixit, A. K., and R. S. Pindyck. 1994. *Investment under Uncertainty.* Princeton, NJ: Princeton University Press.

Fairfield, P. 1994. "P/E, P/B and the Present Value of Future Dividends." *Financial Analysts Journal* 50, no. 4 (July–August): 23.

Ganguin, B., and J. Bilardello. 2005. *Fundamentals of Corporate Credit Analysis.* New York: McGraw-Hill.

Gigerenzer, G. 2014. *Risk Savvy: How to Make Good Decisions.* New York: Penguin Books.

Graffy, E., and S. Kihm. 2014. "Does Disruptive Competition Mean a Death Spiral for Electric Utilities?" *Energy Law Journal* 35, no. 1: 1–44.

Guthrie, G. 2006. "Regulating Infrastructure: The Impact of Risk and Investment." *Journal of Economic Literature* 44, no. 4: 925–72.

Hicks, J. R. 1946. *Value and Capital.* Oxford: Clarendon Press.

Hierzenberger, M. 2010. *Price Regulation and Risk: The Impact of Regulation System Shifts on Risk.* London: Springer.

Higgins, R. C. 1988. *Analysis for Financial Management.* Homewood, IL: Richard D. Irwin.

Holton, G. 2004. "Defining Risk." *Financial Analysts Journal.* Charlottesville, VA: CFA Institute.

Hoppock, D., and D. Echeverri. 2013. "Risk-Based Decision Making: Utility Regulation and Planning under Uncertainty." Presentation to NARUC and SEARUC. June 9.

Jones, C. 2010. *Investments: Analysis and Management.* Hoboken, NJ: John Wiley and Sons.

Joskow, P. L. 2008. "Incentive Regulation and Its Application to Electricity Networks." *Review of*

Network Economics 7, no. 4.

Joskow, P. L., and R. Schmalensee. 1986. "Incentive Regulation for Electric Utilities." *Yale Journal on Regulation* 4: 1–49.

Kahn, A. E. 1988. *The Economics of Regulation*. New York: John Wiley and Sons.

Kahneman, D., and A. Tversky, eds. 2000. *Choices, Values, and Frames*. Cambridge: Cambridge University Press.

Kihm, S. G. 1991. "Why Utility Stockholders Don't Need Financial Incentives to Support Demand-Side Management." *Electricity Journal* 4, no. 5: 28–35.

———. 2003. "How Improper Risk Assessment Leads to Overstatement of Required Returns for Utility Stocks." *NRRI Journal of Applied Regulation*.

———. 2007. "The Proper Role of the Cost-of-Equity Concept in Pragmatic Utility Regulation." *The Electricity Journal* 20, no. 10: 26–34.

———. 2009. "When Utility Revenue Decoupling Will Work . . . and When It Won't." *Electricity Journal* 22, no. 8: 19–28.

———. 2011. "Rethinking ROE." *Public Utilities Fortnightly* 149, no. 8: 16–21.

Kind, P. 2013. *Disruptive Challenges: Financial Implications and Strategic Responses to a Changing Retail Electric Business*. Washington, DC: Edison Electric Institute.

Kolbe, A. L., J. Read, and G. Hall. 1984. *The Cost of Capital: Estimating the Rate of Return for Public Utilities*. Cambridge, MA: MIT Press.

Koller, T., M. Goedhart, and D. Wessells. 2010. *Measuring and Managing the Value of Companies*. Hoboken, NJ: John Wiley and Sons.

Knight, F. 1921. *Risk, Uncertainty, and Profit*. New York: Houghton Mifflin Co.

Lowry, M. N., M. Makos, and G. Waschbusch. 2013. *Alternative Regulation for Evolving Utility Challenges: An Updated Survey*. Washington, DC: Edison Electric Institute.

Lyon, T. P. 2007. "Why Rate-of-Return Adders Are Unlikely to Increase Transmission Investment." *Electricity Journal* 20, no. 5: 48–55.

Markiewicz, K., N. Rose, and C. Wolfram. 2004. "Do Markets Reduce Costs? Assessing the Impact of Regulatory Restructuring on U.S. Electric Generation Efficiency." NBER Working Paper No. 11001.

Melicher, R., and D. Rush. 1974. "Systematic Risk, Financial Data, and Bond Rating Relationships in a Regulated Industry Environment." *Journal of Finance* 29, no. 2: 537–44.

Modigliani, F., and M. H. Miller. 1958. "The Cost of Capital, Corporation Finance and the Theory of Investment." *American Economic Review* 48, no. 3: 261–97.

Moody's Investors Service. 2013. "Negative Outlook Continues for US Unregulated Power Sector; Regulated Utilities Remain Stable." July 30.

———. 2014a. "Consistency and Predictability of Regulatory Decisions Drive Differences in US

Utility Credit Profiles." July 21.

———. 2014b. "Infrastructure Default and Recovery Rates, 1983–2013." May 12.

———. 2014c. "U.S. EPA Carbon Emission Rules: Draft Proposal Appears Credit Negative for Coal-Fired Plants but Others Will Fare Better." June 3.

Morgan, P. 2013. *A Decade of Decoupling for US Energy Utilities: Rate Impacts, Designs, and Observations.* Graceful Systems, LLC (www.gracefulsystems.com).

Morin, R. 2006. *New Regulatory Finance.* Reston, VA: Public Utilities Reports.

Myers, S. C. 1972. "The Application of Finance Theory to Public Utility Rate Cases." *Bell Journal of Economics and Management Science* 3, no. 1 (Spring): 58–97.

Nichols, N. 1994. "Scientific Management at Merck: An Interview with CFO Judy Lewent." *Harvard Business Review* 72, no. 1: 88–99.

Norton, S. 1985. "Regulation and Systematic Risk: The Case of Electric Utilities." *Journal of Law and Economics* 28, no. 3: 671–86.

Peltzman, S. 1976. "Toward a More General Theory of Regulation." *Journal of Law and Economics* 19: 211–40.

Phillips, C. F. 1993. *Regulation of Public Utilities: Theory and Practice.* Arlington, VA: Public Utilities Reports.

Porter, M. 1998. *Competitive Strategy Techniques for Analyzing Industries and Competitors.* New York: Free Press, 1998.

Porter, M. E., and C. van der Linde. 1995. "Toward a New Conception of the Environment-Competitiveness Relationship." *Journal of Economic Perspectives* 9, no. 4: 97–118.

Pratt, S., and R. Grabowski. 2010. *Cost of Capital: Applications and Examples.* Hoboken, NJ: John Wiley and Sons.

PricewaterhouseCoopers. 2008. *A Practical Guide to Risk Assessment: How Principles-Based Risk Assessment Enables Organizations to Take the Right Risks* (www.pwc.com).

Regulatory Research Associates. 2012. *Regulatory Focus*, January 19.

Richards, C. 2013. "Overcoming Aversion to Loss." *New York Times*, December 9.

Rose, K. 1996. *An Economic and Legal Perspective on Electric Utility Transition Costs.* Columbus, OH: National Regulatory Research Institute, Ohio State University.

Ross, S. 1993. "Is Beta Useful?" In *The CAPM Controversy: Policy and Strategy Implications for Investment Management.* Association for Investment Management Research Conference Proceedings, October.

Samuelson, P. 1994. "The Long-Term Case for Equities and How It Can Be Oversold." *Journal of Portfolio Management* 21, no. 1: 15–24.

Shepherd, W. G. 1992. *Regulation and Efficiency: A Reappraisal of Research and Policies.* Columbus, OH: National Regulatory Research Institute, Ohio State University.

SNL Financial. 2013. "Power, Gas Utility CEO Compensation Nudges Up in 2012." May (www. snl.com).

Standard & Poor's. 2012a. "Water: The Most Valuable Liquid Asset." March 7 (www. standardandpoors.com).

———. 2012b. "Standard & Poor's Revises Its U.S. Utility Regulatory Assessments." December 28 (www.researchandmarkets.com).

———. 2014. *Global Infrastructure: How to Fill a $500 Billion Hole.* New York: McGraw Hill Financial.

Taleb, N. M. 2010. *The Black Swan.* New York: Random House.

Train, K. E. 1991. *Optimal Regulation.* Cambridge, MA: MIT Press.

Wein, H. 1968. "Fair Rate of Return and Incentives—Some General Considerations." In *Performance under Regulation*, ed. Harry Trebing. East Lansing, MI: MSU Public Utilities Studies.

Wharton, J., B. Villadsen, and H. Bishop. 2013. "Alternative Regulation and Ratemaking Approaches for Water Companies: Supporting the Capital Investment Needs of the 21st Century." Report prepared for National Association of Water Companies. September.

Wobbrock, R. 2014. "Moody's View of US Regulation." Presentation at the 2014 SURFA Financial Forum in Indianapolis, IN.